I0159997

TRANSFORM
THROUGH
TRAVEL

CONNECTING
ACROSS CULTURES

ROBERT MAISEL

TRANSFORM THROUGH TRAVEL

First published in 2021 by

Panoma Press Ltd
48 St Vincent Drive, St Albans, Herts, AL1 5SJ, UK
info@panomapress.com
www.panomapress.com

Book layout by Neil Coe.

978-1-784529-47-5

The right of Robert Maisel to be identified as the author of this work has been asserted in accordance with sections 77 and 78 of the Copyright, Designs and Patents Act 1988.

A CIP catalogue record for this book is available from the British Library.

All rights reserved. No part of this book may be reproduced in any material form (including photocopying or storing in any medium by electronic means and whether or not transiently or incidentally to some other use of this publication) without the written permission of the copyright holder except in accordance with the provisions of the Copyright, Designs and Patents Act 1988. Applications for the copyright holder's written permission to reproduce any part of this publication should be addressed to the publishers.

This book is available online and in bookstores.

Copyright 2021 Robert Maisel

Dedication

This book is dedicated to my family, friends, and colleagues who have been so supportive and helpful. This book is also dedicated to all of my readers – I hope it helps you to better understand the important impact of travel and connection, and transforms your life in the most positive of ways!

Testimonials

"Robert's insights and perspectives on transformation through travel are both practical and motivational. Just five pages in and I was ready to hop on a plane in search of my next adventure!"

Nicole Barile, Founder, NB Intercultural

"Imagine being in a dark warehouse and having only a weak flashlight to navigate with. Then, imagine that same warehouse with all the lights turned on. It would be a completely different experience, wouldn't it? Travel can illuminate your world in the same way. This book will inspire you to want to turn those lights on. And, once illuminated, your world will never be the same."

Bob Sager, Founder of SpearPoint Solutions

"Transform Through Travel will teach you a lot more than how to just connect with others; it will teach you how to transform yourself, your family as well as your company culture. Highly recommended!"

Falguni Katira, VP, Multichannel Marketing, Transformation Coach

Acknowledgements

One of the most important lessons I've learned in life is that we all need help sometimes. The creation of this book certainly wouldn't have been possible without the help of others. I'd like to take the time to thank everyone who contributed in any way, shape, or form.

Thank you Mindy Gibbins-Klein, for "bringing the book out of me." I am grateful to have been afforded the opportunity to join your group coaching course, which served as an impetus to getting this book written. This course held me accountable and made sure that I followed through with finishing the book.

A special thank you to the Panoma Press Team for your hard work and guidance in bringing these ideas to life by publishing my book.

Thank you Lee Constantine, for working with me tirelessly and for being so patient, even when I pushed back deadlines. Your support leading up to and during the pre-sale campaign was extremely helpful.

Thank you to my photographer, Cliff Toy, for spending so much time and energy capturing and developing a series of phenomenal photos. Your knowledge, patience, and contagious enthusiasm created a relaxing and fun atmosphere, allowing my authentic self to shine through in the pictures.

Thank you to all of the people who read an early version of the manuscript and provided valuable feedback: Lincoln Athas, Jane Kayantas, Nicos Hadjicostis, Jessica Jaroff, Francine Dreste, Selina Sears, Carrie Takamatsu, and Mom.

Thank you to all of my friends and family for believing in me. Believing in someone is one of the most powerful gifts you can ever give them.

Thank you Dad, for showing me the value in making strong connections with others, and the value in connecting those people with one another. By following in your footsteps, a MAJOR theme of this book -- CONNECTION -- unfolded.

Thank you Jim Rothenberg, for being such a loyal friend to my father, and for teaching me many important life lessons.

Thank you Harold Ginsburg, for the care, concern, and support you provided my dad with throughout the years, and continue to provide me with.

Thank you to my stepfather, Kevin Golden, for helping me become a better decision-maker and take responsibility for my life. Through your guidance, I've developed into a more fun-loving and outgoing individual, and have learned the importance of being myself. You showed me the value in taking risks, boosted my self-confidence, and inspired me to work hard and to be proud of what I do.

Thank you to another author, my mom, Gail Maisel, who generously and selflessly poured hours of her time into numerous rounds of edits and coaching sessions that kept me moving forward. Your opinions and sage guidance were invaluable.

Thank you to all of the amazing people I have met during my world travels. You have made an impact on me, have shaped the person I am today, and have contributed to the anecdotes shared in this book.

Lastly, thank you to all of my readers. I am humbled that you have decided to spend your valuable time reading my stories and insights. I am grateful to each and every one of you!

Contents

Introduction

So, you've always wanted to do it. You've seen and heard amazing stories about people traveling the world, and you want to be one of those people. Well, I have good news for you. YOU CAN BE ONE OF THOSE PEOPLE! It's simple, not easy.

My mission for this book is to provide you with the information and confidence to take your dream trip -- whenever you can, be it long or short, near or far -- and to show you, through personal examples from my journeys around the world, how you can transform and connect through travel.

Perhaps it's always seemed impossible to you. That's because it was. Because you believed it was impossible. Everything is impossible until it's been done and therefore proven to be possible.

Have you ever stopped yourself from doing something you really wanted to do because you felt like it was unrealistic or because you didn't believe you could do it? Let that thought linger in your mind as you read on...

I will describe lessons I learned through years of travel as I chose to follow my passion for understanding more about people and cultures. I have had the opportunity to immerse myself in Buenos Aires, Argentina for a semester of university, and in Tokyo, Japan for three years of work. I have been extremely fortunate to have had the opportunity to travel to over 50 countries and to have learned five languages -- English, Spanish, Portuguese, Japanese, and conversational Mandarin Chinese along the way. I've met thousands of incredible people and have experienced deep personal and professional growth. These experiences were so powerful, life-changing, and transformational for me that I felt compelled to share them with you. My goal is that through this book, you will be inspired to find a way to incorporate travel into your life and to find ways to have similar, life-altering experiences.

CHAPTER 1:

WHY TRAVEL?

Well, lucky for you, I'm going to spend the rest of this book answering that question! However, before really getting into it, I figured I'd give you a brief overview of what's in store for you.

Travel has changed my life in *so many* incredible ways! This book will show you how and how it can change yours too!

First, and **OF MAJOR IMPORTANCE, IT'S A HUGE INVESTMENT IN YOUR FUTURE**. I can confidently say that I have gained more practical knowledge from my *Outside-of-the-Classroom Learning Experiences* (aka my travels) than I have from all of my studies and work experiences **COMBINED**. I believe that formal education is a wonderful thing, and I have learned a lot from schools and universities. The same goes for my experiences in the workplace. However, I have picked up information more quickly and have retained it better in real-world settings. This mentality has allowed me to learn about topics I am interested in by selecting destinations where I can have experiences that will support what I am aiming to learn. Studies have shown that when people are engaged and interested in what they are learning, the information is more likely to stick. I also found that what I have learned from the world has been more impactful, meaningful, beneficial, and practical to my life than any other form of learning I have engaged in.

Travel moves the self-development needle in so many ways. It is difficult to capture all of them, however I am going to do my best to do so anyway. The following list is a myriad of ways that travel changes you. We will go on to explore many of these concepts throughout the course of this book:

- Travel makes you more curious.

- Travel makes you more caring.

- Travel makes you more responsible.

- Travel teaches you to manage your finances better.

- Travel teaches you how to become more organized and how to be a better planner.

- Travel teaches you more about yourself than you ever thought or realized you would learn.

- Travel makes you a more creative person.

- Travel makes you more thoughtful.

- Travel makes you a more resourceful person.

- Travel teaches you to be more emotionally mature.

- Travel teaches you to practice self-compassion and self-love.

- Travel makes you a more grateful person.

- Travel makes you more resilient.

- Travel teaches you how to think on your feet.

- Travel teaches you how to be a leader.

- Travel makes you a better communicator.

- Travel makes you a better problem solver.

- Travel teaches you how to manage stress better.

- Travel makes you more independent.

- Travel allows you to take in, analyze, and utilize new perspectives.

- Travel allows you to better understand and evaluate the bigger picture of your life.

Finally, and of MAJOR importance, **TRAVEL ALLOWS YOU TO BE AND FEEL MORE CONNECTED.** Travel has the ability to connect you with the world around you on a physical, emotional, and spiritual level. Travel has the ability to connect you with nature. Travel has the ability to connect you with others like you. Travel has the ability to connect you with others UNLIKE you. Travel has the ability to connect you with yourself.

Now, are you ready to dive in and learn how travel can do all of that? Okay, LET'S GO!!

CHAPTER 2:

YOU CAN DO IT!

PRESSURES, CHALLENGES, AND UNEXPECTED SUPPORT

Everyone can do it. So if everyone can do it, why is it that everyone *doesn't* do it?

While travel is a common pastime, I do not believe that at this point in time, the world universally recognizes travel as an effective means of transforming lives. This is likely because the transformational effects of travel are not fully understood yet, especially by those who have not traveled much or at all. While there is nothing wrong with traveling to relax, to see certain destinations, for business, and so on, there is a bigger piece of the puzzle that is missing in the world's overall outlook on the topic. This book will serve as an agent for change, helping people worldwide to realize the extraordinarily powerful effects that travel has on individuals, communities, and the entire planet.

Benefit from the positive aspects of your society, but don't let your dreams go unfulfilled if they are not easily accepted by others. Every society, whether on purpose or not, molds the people within it. Most people conform, accept certain standard beliefs and ideals, and don't even realize how deeply ingrained these beliefs are within them. These beliefs, coupled with the ways in which we communicate and fulfill our needs, can be defined as culture.

Culture is an amazing phenomenon, one that promotes the creation of relationships through similar values, binds people and societies together, and is therefore an important element of the survival of the human race.

That does not mean however, that you will or must agree with ALL ASPECTS of your culture or society. There may be some or many parts of it that you do not agree with or do not believe in. That's alright. Sometimes we need to **IGNORE** some aspects because they aren't always helpful to follow. Many years ago, slavery was legal and tolerated in The United States. And women couldn't vote. Societal values evolve over time. And it often takes societies a long time to realize that some of their values are unethical and need to change.

Actually, it's **YOU** who knows best about YOURSELF. You know what decisions to make that will be best for your own life. Stepping outside of societal norms and your comfort zone isn't easy. It's a huge risk. But it's a risk worth taking. **In fact, it is more dangerous to AVOID taking this risk than it is to take it.** Read that last sentence again.

Don't let people drag you down. People will tell you that you can't or shouldn't do things, go places, and how impractical your dreams are. But you mustn't let people stop you in your tracks. Don't forget that **almost everyone has an opinion about almost everything**. Rather than seeing this as a nuisance or a pain, let's reframe our thinking and see the good in this situation. Don't let this hurt you; let it guide you. Use people's words, questions, opinions, and judgments -- however harsh they may seem at times -- to refine your thinking and plans rather than letting these things thwart them.

This happened to me on numerous occasions, from just about every friend and family member I have. Actually, being challenged is a good thing. It forces you to think critically, carefully outline your goals, and make decisions that align with them. People pass judgment and criticize for a variety of reasons. They may be concerned for your safety and wellbeing. They may want you to take a different path, maybe one that entails less risk. Or they may be scared of judgment, failure, and going against societal norms themselves. They might be jealous that you're planning on following your dreams, perhaps doing something they have always wanted to do but for whatever reason haven't. They therefore may project those negative emotions onto you, oftentimes unwittingly. Please identify these people in your life and do them a favor -- provide them with a copy of this book -- they need to read it too.

And speaking of you, **YOU** can be, and often are, **YOUR OWN WORST ENEMY. Don't let YOURSELF hold you back**. If you know a path is right for you, THEN IT IS. Trust your gut feeling. **Have the courage to follow the path less traveled.** Have the wisdom to know you're on the right path, even if nobody around you sees it. Or understands it. Or supports it.

During my 14.5-month trip throughout Asia, my cousin sent me an email. "Be at the Felix Bar in the Peninsula Hotel in Hong Kong at 7:15 pm sharp. You have a dinner reservation. And it's already paid for. Oh, and don't forget to visit the bathroom. Even if you don't need to use it." I was in shock, so thankful for this opportunity! 'But why should I visit the restroom...especially if I don't need to use it?' I pondered...

I arrived before the time of my reservation. The building oozed with luxury. Again, the bathroom comment entered my mind. 'I wonder if they have really nice toilets...' I thought to myself.

Through the lobby and into the elevator I went. The Felix Bar was on one of the highest floors of this building. This wasn't exactly pleasant news as I am fearful of heights. Oh well, there was no turning back at that point; I was already in the elevator.

The door opened, and I walked into one of the finest and fanciest spots I'd ever seen. Ornately arranged and decorated, the dining room was impressive to say the least. Very impressive. And over to the window I didn't want to go. But over to the window I went. And I looked down. And it was a long, long way down. The view was dizzying. But at least from up here I had a stellar view of the city! The view couldn't get any better! ...or so I thought...

When I sat down to eat, I was presented with an extraordinary menu. It was hard to select from all of the delicious options. Once I made the difficult decision of what to order, I could only wait in anticipation of the food. And once it came, I was anything but disappointed. I proceeded to have one of the best meals of my life! I was so grateful to my cousin for arranging this for me.

And then it hit me. The final thing my cousin had said. To go to the bathroom. Surely, I couldn't leave this bar without following through on such a simple yet strange request...so I decided to check it out.

I opened the door and found a marvelous combination of grandeur and beauty wrapped in one. The bathroom was nice. REALLY nice. Fancy. The kind of fancy restroom you would expect a fancy bar to have. The toilets and sinks were lavish. I could see why my cousin would want me to have a look at this for myself. Clearly, he appreciates the restrooms of fine establishments, and perhaps even more so than I had realized.

And then it hit me, throughout life, I had been so busy focusing on the food at restaurants that I had failed to appreciate such an important element of the equation -- the restrooms! Here I was thinking only of the main reason I was there when I should have been reveling in the entirety of the experience! Oh, how closed-minded I had been up until this point, failing to truly understand the beauty and vital role these rooms play. Surely I would leave this establishment more appreciative of the décor and importance of restrooms than ever before!

And then it REALLY hit me. What my cousin was ACTUALLY talking about. The reason I was TRULY standing in the bathroom of a bar so high up in a hotel in Shanghai.

As I locked eyes with the urinals, and the urinals locked eyes with me (okay, I can't prove that urinals actually looked at me, but it's highly likely, right?), a beautiful sight appeared! No, not the urinal itself, but the backdrop. Most urinals are placed in front of a wall. These urinals were not.

What lay behind these urinals was a window. A window the size of a wall. A window INSTEAD of a wall. 'Why would anyone want to pee in front of a window? Isn't that an invasion of privacy?' I thought. Positioning oneself in front of the urinal clearly allowed for an insanely breathtaking and relieving experience all at once. Both for the reliever and for those on the street looking up at the bar's restroom window.

For the reliever, the view offered was an even finer angle of the city than what is afforded by the windows in the dining section of the bar.

And for those on the street looking up, a relieving view of staring at windows that do not provide any clue as to what's behind them. Thank goodness for one-way glass.

And oh what a glorious recommendation from my cousin it was. Profusely thanked he was. And forever grateful I was for the opportunity to have such a marvelous experience!

As you saw from the previous story, people will support you in ways that will surprise you. Not just people you know, but people you don't know as well. A prime example of this was during a road trip that my friend Lincoln Athas and I took. On a stop at the local mechanic to assess an issue the car was having, we explained that we were on an epic 5-week adventure around The United States to learn more about its history, experience the nature and beauty of our country, meet interesting people, have unique experiences, strengthen our friendship, learn more about ourselves in the process, and so much more. The mechanic was so impressed and inspired by what he had heard that not only did he resolve the issue and get us up and running that same day, he also refused to take even a dime in compensation for the parts or the labor. This act of kindness nearly brought me to tears. I hadn't seen that coming. Neither had Lincoln. But neither had the mechanic. This mechanic had NEVER in his life encountered two souls brave enough to seek out and embark on this kind of a mission.

This mechanic stop would prove to be one of eight. Oddly enough, the car, which had never given Lincoln any trouble in the past, was acting up in unimaginable ways. With a rearview mirror left dangling by a wire, to the trunk being locked shut (with our belongings conveniently located inside of it), and everything else you can think of and more having happened in between, to report that this wasn't an adventure in every sense of the word would be pure and utter nonsense.

Through these experiences, I began to wonder, 'Has the universe sent these challenges our way to see how we respond? To teach us lessons?' Our responses showed how driven (literally) we were to succeed, and it felt as though the universe responded accordingly. I have come to believe that when someone wants something badly enough, the universe tends to sense that and helps the person to attain it. Perhaps if you want something badly enough in your life, the universe will sense that and will assist you too.

Lincoln and I were challenged in many ways on this journey. We could have given up and headed home at many points. But we didn't. Instead, we embraced these challenges head-on with a burning passion to succeed, coupled with eager, jovial, and lighthearted attitudes. With an average driving time of six hours per day, visits to 32 states, learning about history, culture, ourselves, and so much more in the process, we achieved our goal of road-tripping extensively throughout The United States in a methodical and calculated fashion. Being in the cities we wanted to be in for the weekends. Adjusting our plans to avoid inclement weather. And all within just five weeks. By completing this epic adventure, we proved that we could accomplish ANYTHING we set out to do. We accomplished what would have been seen by most as impossible. And proved it to be possible. **Remember, with the right mindset, we are all UNSTOPPABLE!**

FEAR AND HOW TO OVERCOME IT

Fear. Ever heard of it? Ever experienced it?

This word holds us back from SO many things. DON'T LET IT. You have a choice. Doing something that scares us (unless of course it is blatantly dangerous) is one of the most powerful ways to learn, grow, and achieve. How do I know? Because I have stared fear in the eyes and have backed down many a time. And that's how I learned NOT to back down. I have learned to stare fear in the eyes, acknowledge its presence, and politely ask it to step out of my way. I have big things to accomplish in this life, and so do you.

The source of fear is usually the unknown, so let's counter ignorance by learning and trying new things!

How many things have you NOT done in your life because you were scared? How many opportunities did you let slip away because you were afraid to try something new?

Fear not, and don't dwell on the past. Instead, let's learn from it, alter how we live in the present, and plan for the future accordingly!

Fear can only hold us back if we let it. Only if we become its prisoner. Many people don't realize this, but **WE HAVE THE POWER TO MAKE THAT CHOICE.** We don't need to stay locked up in that jail. We hold the key to the cell and have the ability to unlock it at any time. **Instead of letting fear paralyze us, let's use it as a motivator to propel us forward to the places (both literally and figuratively) that we've always dreamed of being in life!**

I still recall staring at the screen, looking at the one-way flight from New York City to Beijing, China that I was about to purchase... holding my right index finger down on the left-click button of the mouse and feeling extremely nervous. I was scared that I might be making the wrong decision. It was the summer directly after completing my MBA. I should have been out looking for work like the rest of my diligent classmates, not on the internet about to purchase a ticket to a far-off land I had only ever dreamed about, right? I had to be crazy. For sure, I was making the wrong decision... Actually, *are you sure about that...?*

I'm sure of one thing, and that is this: **I ABSOLUTELY MADE**

THE RIGHT DECISION.

Society certainly didn't support my decision. My family wasn't happy about it. My friends couldn't understand it. But I SUPPORTED MY DECISION. I was happy with it. I understood it. I knew that I needed to take that trip and that somehow, it would fit into the grander scheme of my life and the impact I would make on this world. And that's all that mattered.

Letting go of the left-click button of the mouse, watching the icon spin in the middle of the screen, my heart skipped several beats and the butterflies in my stomach began dancing even more wildly. The screen changed to show a confirmation page. It was over. My decision had been made. I was going. My life was about to change drastically. For the better. And as you continue reading, you'll learn not only how my life changed through travel, but how yours can too.

DOUBT AND HOW TO DESTROY IT

Fear leads to doubt. At best, doubt clouds our vision and acts as a negative force standing in front of us and our dreams. At worst, it stops us from going after our goals and dreams entirely. We must realize this and remove doubt from our minds.

So how do we counteract doubt? How do we climb over this seemingly insurmountable mountain that seems to block our grand plans? I'm not asking this question rhetorically. I'm going to answer it.

The answer is self-confidence. Believing in yourself. **Every human on this Earth is AMAZING and can accomplish MIRACULOUS things**. It's just a matter of understanding that.

And TRULY BELIEVING that.

People worry about what others will think of them if they do certain things. People worry about what others will think of them if they DON'T do certain things. People doubt themselves and their abilities. **EVERY. SINGLE. DAY.** How do I know? **Because I was one of those people**.

People become consumed with what society will think about their story if it gets out. People worry that they haven't accomplished much in life. Ironically, people actually tend to be more afraid of success than they are of failure.

WE need to believe that our story is amazing before **OTHERS** can and will believe that it's amazing. Have you ever tried to sell something you're not interested in? You probably had no luck making sales. Why? Because there was no passion behind your words, gestures, and actions. It was likely clear to the other party that your interest in the product or service wasn't genuine. The same effect applies to our belief in ourselves, our stories, and our lives. If we don't authentically believe in the power of our stories, others won't either.

I have found self-belief to be the precursor of and strongly tied to confidence. When I have believed in myself, I have achieved what I set out to do. And when I have *not* believed in myself, I have failed. Based on my own experiences, I would propose that believing in ourselves is a prerequisite of success. That last sentence is powerful; please re-read it.

My experience has shown that our attitude also plays a *crucial* role in influencing our outcomes. To experience GAMECHANGING RESULTS, we need to have a GAMECHANGING attitude. And the right attitude comes from having the right mindset.

Do you believe in yourself? Is there anything you're really interested in doing but are worried that you might not be able to do it? These are completely normal thoughts to have, but we have to realize that they are just thoughts. They are not our reality *unless we let them become our reality*. Let me say that again – THEY ARE NOT OUR REALITY UNLESS WE LET THEM BECOME OUR REALITY. This is OUR choice. WE create the reality we want in our lives. Therefore, let's make our minds fertile to thoughts of encouragement, achievement, and success, and allow those thoughts to dominate our thinking. Let's work to turn THOSE thoughts into reality.

I'm a normal person just like you. I've been EXTREMELY scared to do many things in my life. It's been a matter of believing in myself and knowing that I am capable of achieving great things, just as you are, which has led me to push past fear and self-doubt and has led me to **TAKE ACTION** and achieve extraordinary results. Any travel-related goals I have set, I have met. And that's because I have believed that I could accomplish them.

I wanted to see the world, but that little voice tried to instill fear in my heart. It told me I couldn't do it. It questioned my decisions pre-, during, and post-trip. On EVERY SINGLE EXTENDED TRIP I HAVE EVER TAKEN. It was not easy to do, but I came to the realization that I DON'T HAVE TO LISTEN TO THAT VOICE AND NEITHER DO YOU. I write this book today, proud to tell you that I managed to push this voice aside and accomplish great things. Yet the voice still comes back quite often in life and tries to dissuade me from doing great things. It still affects me sometimes -- I'm human just like you. As a result of deciding not to listen to that voice, and going after what I knew would be beneficial to myself, others, and the world around me, the following amazing journeys (in chronological order, and all taken alone except for my United States Road Trip) have come to fruition:

- A 2-Month Europe Trip

- A 14.5-Month Asia Trip

- A 3.5-Month South America Trip

- A 5-Week United States of America Road Trip

- A 7-Month World Trip

I used to let my fear of others' thoughts, opinions, and judgments hide my amazing adventures, especially in the professional world. I would try to avoid the topic in job interviews, scared that I would be judged for taking extended amounts of time to travel instead of working. Or I would minimize the impact of these trips when asked about them. Until one day I came to an extremely important realization. I realized that these travels are the core of who I am. That through them, I learned about myself, the world, new languages, and how to better understand cultures. I realized that **THE BIGGEST DISSERVICE I HAD BEEN DOING MYSELF WAS NOT BEING MYSELF**. Re-read that please. After this enlightening realization, I vowed to NEVER be afraid to show my true self to the world again. Some people will accept us for who we are. That's fine. Other people will NOT accept us for who we are. That's fine too. **At the end of the day, what matters is that WE ACCEPT OURSELVES FOR WHO WE ARE.** And this comes from living a life of purpose, true to our values, making decisions based on our beliefs. Without the fear of showing others who we really are. This isn't easy to do, but in my opinion, it's crucial. If I can do it, so can you.

We must realize our true potential and push forward with what we know is right for US. What we feel is right in our hearts and in our "guts." I'm not saying we shouldn't evaluate decisions logically. I'm also not suggesting that we act impulsively, and solely based on our emotions. But our hearts and our "guts" tell us things we may not otherwise realize. There is scientific evidence behind the fact that our "gut feeling" is a way of our body communicating with us -- a "sixth sense," if you will, steering us in the direction we are meant to be going in. We therefore need to listen to our bodies. Have you ever felt your body was trying to tell you something? Do you know what it was trying to tell you? Did you listen to it?

THE IMPORTANCE OF BEING PRESENT

Our brains are on autopilot for the majority of our lives. We aren't usually aware of what we think and do or why unless we take the time to think about our thoughts. And to think about our actions. And to think about why we think and act the way we do. It is therefore important to slow down and make sure we are being present in our lives. This way, we can accurately assess and analyze situations and identify where we may want or need to make changes.

Meditation is a great way to be present and to become aware of our thoughts. By closing our eyes and focusing on our breath or using a mantra, we are able to relax our bodies. Meditation also enables us to be more present, and therefore more aware of our thoughts and sensations. Meditating, even for just a few minutes a day helps to increase focus as well.

While traveling through India, I would go on to learn just how powerful meditation really is...

I made a friend during my travels, and we spent some time in a place called Rishikesh. It was there that I was introduced to yoga and meditation. We found and stayed in an ashram, whereby for staying at least five nights, yoga and meditation classes were included in our stay.

We also explored another ashram in the woods of Rishikesh, away from the touristy part of town. Covered by thick brush, it was quite a trek to get there. The buildings were abandoned. Broken windows. Extremely large spider webs housing extremely large spiders. Gigantic black mosquitoes adorned with white dots. The kind that carry malaria. But it was all worth it. That same ashram was where the Great Maharishi had spent time many years ago. That same ashram was where a world-renowned band spent time meditating and writing songs. That same ashram was where The Beatles wrote many songs that would end up in *The White Album*. Having grown up a raving fan of The Beatles made this experience all the more powerful for me.

My friend and I continued to do yoga and to meditate even after leaving Rishikesh. Daily. And then we went to the outskirts of a city called Jaipur. And we did a 10-day meditation course. A SILENT meditation course. Which meant no talking. At all. Not even a word. For 10 days. In a row.

That 10-day meditation course was one of the most impactful events I have ever participated in throughout the course of my entire life.

Frustrating. Challenging. Painful. Those would be the words I would use to describe the experience *while* I was having it.

Spiritual. Beautiful. Enlightening. Those would be the words I would use to describe the experience *after* having it.

We were in the middle of nature. In the woods. Rustic and serene beauty abound.

We were instructed that we would not be able to speak, that we would watch a video each night, and that we would have access to vegetarian meals.

I remember the man speaking in the videos. The wisdom he shared was enlightening. These videos were fascinating and really made me think. Something I remember vividly is the gentleman on the screen questioning the meaning of possessive pronouns such as "mine," yours," and "ours." He seemed to believe that these words were ridiculous to use, claiming there is no such thing as possession in this world. That we own nothing. And that we are just on this Earth for a limited amount of time, and during that time, we are borrowing the items we claim to and think we own. That was a powerful new perspective to consider, and I liked that it challenged my way of thinking.

The food was very tasty! And sometimes my travel partner and I would sit next to one another during the meals. For support. Not to sneak in a conversation...or so I thought... And on one occasion, he spelled out some words with his food (we weren't supposed to do that either), trying to communicate with me. I started to look over at the message and saw "Wan..." and then looked away as I began to feel guilty, knowing that it was imperative not to communicate with anyone in any way, shape, or form so as not to disrupt the experience for myself or anyone else. So I looked away and didn't finish reading what he had written.

I believed that through the difficulty of this experience, he was looking to provide some humor and had written "Wanna chai?" This had become a fun, inside joke between us since we had both come to love chai throughout our time in India. And we had found that saying it in a certain fun and uplifting tone could brighten our spirits on even the darkest of days. Chai, by the way, is short for masala chai, which is a tea beverage that is made by boiling black tea in milk and water, with a mixture of spices and aromatic herbs added in.

But this was not the message he had written. Not even close. "Want to leave?" was what he had actually written. After the meditation was complete and we were able to speak again, we discussed the matter. He had assumed that I had seen and read the message and that by me not responding with a nod of affirmation to his question, he had believed I wanted to stay.

The truth was that I was suffering. He was suffering. We were all suffering. From our internal thoughts plaguing us day in and day out. From the physical pain of having to sit on the floor for so many hours, for so many days straight. And had I actually read the full message, I may very well have been tempted to leave the program early with him. I am glad I didn't become aware of his desire to leave until after the program was over and that we pushed through and stayed for the entire thing. Not being able to communicate with other humans was hard. Because it wasn't what I was accustomed to. However, by not speaking with anyone, I was able to focus on communicating with both myself and with nature.

I thought a lot. During the meditation sessions and outside of them. Without the ability to communicate with anyone or use technology, there isn't a whole lot more one can do. I learned about the kind of thoughts I was having. I finally had the opportunity to slow down and actually become aware of my thinking (for an extremely extended period of time) instead of letting it go on autopilot as it usually does. This provided a helpful and powerful window into better understanding myself.

And connecting with nature. That was a profound experience too. I remember admiring the trees. The beautiful swarm of butterflies. The warm rays of sunlight shining on my body. The cool wind as it swept across the Earth. Admiring the little things in life. The beautiful things in life. The amazing things in life. The things in life that we have grown so accustomed to ignoring because we live in such a fast-paced world. Full of stimuli. That numbs our awareness of the true beauty that lies directly before our eyes.

This experience taught me that as much as I love to speak, I don't have to. That I can survive without doing so. That it wouldn't be easy, but that it would be possible. I also learned discipline. That I can train myself to act or not act in certain ways. To sit still. To not fidget. To ignore an itch (try doing this, it's harder than you think). To not give up in the face of adversity. And to appreciate. To appreciate my mind. To appreciate my thoughts. To appreciate my body. To appreciate my life. And to appreciate the beauty that I am fortunate enough to be surrounded by on a daily basis.

I am thankful that this friend opened up my eyes to the value and importance of these amazing activities. I have carried both yoga and meditation into my life today. I sometimes use yoga to stretch my body and warm up my muscles (though I am aiming to be more consistent with this). I use meditation to increase focus, raise awareness of my thoughts and sensations, be more present, reduce stress, and connect with my spiritual side.

Active listening is another great way to be present. So often in life, we are distracted when others are speaking to us. We are often either not paying attention at all (likely distracted by technology) or paying the other person only partial attention. We may be thinking about what we are going to say next rather than actually listening to the other person. Active and engaged listening is when we listen with the intent to understand, allowing for our responses to be thoughtful and to reflect what the other person just finished saying. Sometimes even repeating a few words the other person just said. Being an active and engaged listener is a key skill involved in being fully present. It's easier said than done.

Perhaps most importantly, we need to remember to always live in the present. We certainly *can* and *should* acknowledge and learn from the past, but let's not get stuck there. We certainly *can* and *should* think about and plan for the future too, but let's not get stuck there either. Let's not forget about the now. **WE MUST**

REMEMBER TO ENJOY OUR LIVES BY LIVING IN THE MOMENT. If we spend all of our time rehashing the past and planning for the future, we will never find the time to fully embrace and enjoy the present.

THE RIGHT MINDSET

It is my belief that we need to possess and maintain a positive mindset to succeed both in travel and in life.

Anybody who knows me would tell you that I'm an extremely positive thinker. I'm going to drop something on you right now that you probably didn't see coming. I didn't always use to be that way. AT ALL. As a matter of fact, I used to be the opposite. A complainer. Somebody who anyone in their right mind would have wanted to avoid at all costs. This goes to show us that the only constant thing in life is change and that WE have the ability to make a change at ANY POINT IN OUR LIVES. So how did this shift from negative to positive thinking occur?

It was back during elementary school. I lived in Chappaqua, New York at the time and I was about to have a playdate with my friend. My mom was used to the routine. She would tell me that my friend was on his way, and I would complain about him. About the experience. About everything. Until one day, my mom asked me a simple question. "Robert, what do you *like* about this friend?" "Well…" I stammered, "I uh, …well, he's uh…he's…fun to be around…I guess…we play video games together and um… that's…that's fun…" Getting those words out of my mouth wasn't easy. Getting those words out of my mouth was a struggle. Getting those words out of my mouth was vital. Vital in becoming aware of a whole new reality. Vital in experiencing a whole new reality. Vital in believing in and embodying a whole new reality. Vital in beginning the process of becoming a whole new person.

Sure, I could find the negative in situations. I was quite good at that! But what was the point? What would I get out of it? What was the value in thinking this way? How would that affect others around me? Would they want to be around me if I thought and therefore acted this way? Would I even want to be around myself? As I began to ponder these questions, I began to come upon the answers on my own. I began to understand the futility in, and harm caused by negative thinking.

That day, I realized I could view life in a different way. I realized that I HAD A CHOICE TO MAKE EACH AND EVERY DAY. I didn't need to focus on what was bad about my friend and my relationship with him. Instead, I could choose to focus on what was good about him and our relationship.

I then realized that this didn't just apply to this friend and my relationship with him, rather to EVERYTHING IN LIFE. This was an enormous breakthrough! It was that day that I began to realize my ENTIRE LIFE could change RADICALLY, depending on what I chose to focus on and how I chose to view and interact with the world. It was that day that I began to transform from a bratty and entitled child full of resentment and anger toward the world into a more positive, good-natured, and friendly human being. And I found that when I looked for the positive, I found it. And that by focusing on the positive, my happiness and gratitude increased IMMENSELY.

This shift in thinking also allowed me to view challenges as opportunities rather than roadblocks.

I'm not suggesting that we never consider the negative aspects of situations or negative outcomes of decisions – doing that is actually quite important.

What I am suggesting is that we choose to FOCUS OUR ENERGY on the positive rather than the negative. That we bring positive influences into our life and discard negative ones. We must therefore be selective in who and what we allow into our lives and into our minds. What we choose to read and watch affects our thoughts and actions. Let's focus on fun, uplifting, and engaging content rather than that which is disturbing, dreadful, or horrid.

Another reason why ingesting positive information is so important is because intake = output. The information we allow into our minds greatly affects the information we send out to the world. We therefore must take in positive information in order to think positively and to be able to spread positivity to others and to the world that surrounds us.

It's important to realize what we are focusing on, and if it isn't already the positive, to make the shift from negative to positive.

Are you more of a positive thinker or a negative thinker? Did you ever realize that it's **YOUR** choice?

TAKING ACTION

After we have pushed past fear and doubt, are focused on the present and are aware of our thoughts, and have commanded a positive mindset, the final step is to **TAKE ACTION**! I can't stress how important this step is. **We can have the most amazing thoughts, desires, trip plans, and so on, however if we fail to take action on them, they will remain thoughts and ideas, and we will never realize our goals and dreams.** Making a plan for taking action really helps! The steps to success are simple, not easy. Here is a formula that you may find helpful in achieving your goals:

1) Set a specific goal and write it down on a piece of paper

2) Create tasks/routines that will set you up to succeed

3) Hold yourself accountable, or find someone else who will

Have you had any amazing thoughts that you took action on? If so, what happened and how did you feel afterward? Have you had any amazing thoughts that you DIDN'T act on? If so, what was the result and how did you feel afterward?

Going after what you want in life and getting it is INCREDIBLY rewarding. To me, this concept is so important in living a fulfilling life -- one in which you feel that you have value. That you have followed your path. That you have contributed to society. That you have fulfilled your purpose. That you have made a positive impact on the lives of others. That you have left your imprint on the world. That you will leave a legacy behind.

The concept of achieving what you set out to do is empowering. Going after your dreams, surmounting the obstacles in your path, and pushing forward to success feels good. It FEELS RIGHT because it IS RIGHT. It instills in you a sense of PRIDE and ACCOMPLISHMENT that nothing in this world can take away from you. And that nothing else in this world can give you. No book you can read will magically deliver this feeling to you. No picture you look at can provide you with it. You can't exchange goods or money for it. You can't find it within the deepest depths of the ocean or in the furthest corners of space. It must be EARNED. And it must be earned by YOU.

It's YOUR story. You have one shot at life so make it count.

Be proud of your experiences. They encompass who you are. They are YOUR STORY.

What are some of the experiences you are most proud of having? What are some major things you have accomplished in your life? What have you not yet accomplished in life but will in the future? What stories will you tell your grandchildren?

I had plans to travel around Europe with a friend of mine during the summer between the two years of my MBA. I was ecstatic! We would meet amazing people. We would explore amazing cities. We would eat amazing food. We would have a great time!

All of those amazing thoughts came to a screeching halt the day he called me up and told me he wouldn't be able to make it. So much for all of the wonderful adventures I would go on throughout Europe. Sadly, my amazing dream had seemingly just been shattered.

I could have still gone to Europe alone. Or, I could have decided not to go to Europe that summer. I thought to myself, 'It would be scary. It probably wouldn't be any fun without company. I likely wouldn't have an enjoyable experience. And I could always go at a later point in time, right?' The path seemed clear. I knew what I needed to do, and I made the best decision possible with the information available.

I chose to ignore all of the excuses I had created for why I shouldn't go to Europe alone. I decided to set aside the rationalizations my mind produced about why I shouldn't travel that summer. I decided to embark on a journey to Europe. I decided to embark on a journey as a lone warrior. I decided to embark on a journey as a nomad. I decided to embark on a journey that would change my life forever.

It was one of the most difficult yet rewarding choices I have ever made. That decision showed me that I was capable of traveling alone. That trip boosted my confidence to travel alone to such a high level that three subsequent solo trips, much longer and much more involved, ensued because of it.

Because I chose to travel Europe alone that summer, I instilled a MAJOR sense of confidence and independence in myself. I realized that I could meet amazing people. I could explore amazing cities. I could eat amazing food. I just didn't NEED to travel with someone else in order to do these things. And as a matter of fact, by going alone, I was MUCH MORE inclined to meet amazing people, explore amazing cities, and eat amazing food with them. Because I DIDN'T HAVE a travel companion. And that made it easier for me to approach others. And that made it easier for others to approach me.

Sometimes in life, we need to take risks. We need to go out and pave our own path. Taking this unique path is often paramount in realizing our true greatness. It's not easy, but it's necessary, and it's achievable. What's your true path? Will you choose to follow it?

CHAPTER 3:

THE TIME IS NOW (OR AS SOON AS POSSIBLE)!

In the future, you won't want to look back at your life, wishing you had followed your true passions. Realizing you followed the path that was expected of you from others and society. Realizing you didn't live the life you truly wanted to live. This is a powerful paragraph. Take a moment to carefully re-read it. Now re-read it a third time. SLOWLY. IT'S THAT IMPORTANT.

So let's ensure this doesn't happen. Let's make sure we are living a life that aligns with our goals and aspirations. Let's make sure we are living a life that is going in the direction we want it to go in, and if it's not, let's assess the situation to understand why and make the necessary changes. Let's make sure we are living life in such a way that *when we reach our final days, we are content with the life we have lived.*

The reality is that, unless and until science proves otherwise, our lives are finite. We often procrastinate, put things off, and believe that we have forever to do them. When in reality we don't. NONE of us do. Coming to grips with our own mortality isn't an easy concept. It isn't generally one that is brought up at the dinner table. Or talked about on a first date. Or a second. Or a third. While it's not a rosy topic, it is CERTAINLY one we must consider.

Over the years, I have seen family and friends fall ill and pass away. My father succumbed to Alzheimer's Disease. My mother battled cancer (but luckily survived). A high school friend of mine died of brain cancer. A college friend of mine was killed during a moving accident. I had a near-death experience (and you will read about it later in this book). These experiences have instilled a sense of urgency in me. Perhaps now you can better understand why I REFUSE to put off my goals and dreams. And why I believe that you should REFUSE to put off YOUR goals and dreams.

If you died tomorrow, would you be able to say you've achieved something meaningful? Would you be able to say you've achieved all of your life goals? Some of them? Would you be able to say that you lived a life of value and purpose? If your answer to any of these questions is "no," then you've got some thinking to do and some actions to start taking.

You can begin planning your dream trip, or you can take other actions such as beginning to think about and do research for it. If we take no action, our dreams will stay dreams forever and will never materialize. However, if we break them down into specific, actionable, and measurable goals and back them with routines to hold ourselves accountable, we'll be in a much better position to fulfill our dreams!

If you don't start now, you'll procrastinate. It's the story of many people's lives. Has this ever happened to you? Have you become complacent about something you really wanted to do, put it off, and ultimately ended up not doing it?

Excuses. Ah yes, excuses. We as human beings are extremely good at making them. Yet they harm us in so many ways. Not only can they interfere with our daily and weekly productivity, but they can cause us to put off developing our best ideas. When was the last time you made an excuse for something you really wanted to do and never ended up doing it?

"Life" is a popular excuse. Life will ALWAYS get in the way. If you let it. So DON'T LET IT.

If you wanted to start generating ideas for your dream trip right now, could you? Well, I'd like you to think about some things in your life that are real and must be considered.

Let's think about your situation. Let's dive deep and analyze it. This will help you to identify your responsibilities. After you finish reading this paragraph, you will be barraged with an intense amount of questions. But don't worry. You'll survive. And after thinking through these questions, you should have a better understanding of your situation, if/when you may be able to travel, and if so, for how long:

LIVING SITUATION: Are you living with your parents, roommates, or alone? Are you renting a place, have a place of your own, are camping out in your parents' basement, or something even wilder than that?

Exercise: Make a list of all of the commitments you have related to your living situation. Could you alter anything? Could you find someone to sublet your spot in the apartment for a few weeks, months, or years while you're away, or have someone move into your home/a portion of your home and pay rent to cover all or part of your mortgage? Could you partake in a home swap or timeshare situation?

CAREER: What kind of job do you have? Do you work for a company? For yourself? Both? Do you work in an office or remotely? How much time off do you get each year? Could you take it all at once?

Exercise: Make a list of all of your work-related commitments. Could you change anything? Could you work remotely? Would you have to quit your job to take off on your ideal journey? Could you plan a trip within the boundary of your allotted vacation time?

RELATIONSHIPS: Are you single? Married? In a relationship? Seeing someone? If you're in a relationship, is it serious? If so, would this person want to travel with you? Where would this person want to go? If this person did not want to go, would he/she be alright with you traveling for a while without them? If so, for how long?

FAMILY: Do you have a family? Do you have children? If so, how many? Do you live with and/or take care of anyone who is sick or elderly?

Exercise: Make a list of all of the commitments you have to the people in your life. Could you and/or they afford to travel for some time? If so, for how long? Could you make any adjustments to allow for the travel to take place? Could you make any compromises with any or all of these people to allow you to take the trip you've always wanted to take?

FINANCIAL: How much money would you need to make your dream trip happen? How much money do you have saved? If you don't yet have enough money put away, how could you obtain it? How long would it take you to save up enough to take that dream trip? What do your financial commitments look like? Are you paying off student loans? Are you paying off a credit card or seven?

Exercise: Make a list of your savings, assets, and financial commitments. Also, include the amount of time you would like to be gone for. Factor in how many people will be going on this trip, what kind of accommodations you'll stay in, what kind of activities you'll do, and how much you think you'll spend on food, souvenirs, etc. If you're not sure how much you'll spend on these items, you can research their average costs for the region you will be visiting to gain a better understanding. Considering all factors involved in your trip, make sure to **include an emergency budget for unexpected costs.** Unexpected things will almost certainly occur, especially if it's a flexible trip. Trust me.

EXTERNAL FACTORS: Is there anything outside of your control happening? Is there a pandemic going on? Has the region you'd like to visit recently been hit by a devastating hurricane? Did an earthquake cause massive damage to the city you're aiming to spend a few weeks in?

Exercise: Make a list of all extenuating circumstances that you are presently aware of. All factors that are outside of your control. This will be helpful in assessing whether the appropriate time to travel will be soon or if it will be further along in the future. **While it is good to recognize these items, it is futile to expend mental energy worrying about or focusing on them since you aren't able to influence them.**

So now you've assessed your situation and figured out a way to make this happen! IF YOU CAN TRAVEL NOW (OR SOON), DO IT

BECAUSE YOU NEVER KNOW WHEN YOUR SITUATION MAY CHANGE. Let me repeat that! **IF YOU CAN TRAVEL NOW (OR SOON), DO IT BECAUSE YOU NEVER KNOW WHEN YOUR SITUATION MAY CHANGE.** However, if your present circumstances don't allow you to travel right now, don't fret. If you're in that situation, then now is the perfect time to BEGIN THINKING ABOUT, RESEARCHING, AND PLANNING YOUR TRIP! AND YOU CAN TAKE THE TRIP ONCE YOU'RE ABLE TO!

Not having accomplished our goals and realized our dreams just yet isn't something to feel bad about, rather something to look forward to! We just need to remember that the time to start thinking, planning, and taking action is NOW (OR AS SOON AS WE ARE ABLE TO)!

We don't have forever to turn our dreams into reality, and we don't have a magic crystal ball that tells us how much time we have left on this Earth. We must act as soon as we can because our situation could change at any point in time. We could be diagnosed with a deadly illness. And learn that we have three months left to live. Or we could get into a tragic accident tomorrow. And become paralyzed. Or lose our life. **AND THAT IS PRECISELY WHY THE TIME TO ACT IS NOW (OR AS SOON AS WE CAN)!**

CHAPTER 4:

ARE YOU CURIOUS?

Are you curious? Curious about all the new places you'll discover? Curious about all the new people you'll meet? Curious about the new experiences you'll have? These are some of the reasons I love travel so much! There is a whole world that awaits you, and I cannot wait for you to discover it!

ARE YOU CURIOUS ABOUT PLACES?

The amazing places you will see. Oh yes, the amazing places you will see. I can still clearly remember spending several nights watching epic sunsets unfold in the hammock of my personal bungalow overlooking the riverine archipelago of Si Phan Don in Cambodia. For five US dollars per night. You read that correctly -- five US dollars per night. I can clearly remember the incredible view of karst mountains from a rooftop in Yangshuo, China and cycling through the town to explore its many treasures and wonders. I recall some of the most amazing nature and wildlife in the plains of Maasai Mara in Kenya.

The pyramids of Egypt are just as ancient, yet more mysterious in person than they are in pictures. It's one of those "you need to see it to believe it" type of locations.

Many people have told me that they don't physically need to visit a destination. They are content with gaining information about it through a television documentary, seeing photos of the place, or learning about it in some other form. While one can certainly gain an appreciation for travel, cultures, historical sites, and so on from these methods, it is most certainly not the same as actually going to these places. Being there IN PERSON is an ENTIRELY DIFFERENT EXPERIENCE and one that will be unique to YOU. When you visit the place, you LIVE the place. Read the previous sentence again.

With photos, documentaries, etc. you lack the smells and aromas in the air, the taste of the local foods, the sounds of other languages, the interactions with others involved in the experience with you, and the feeling you get inside when you see something as epic as the ancient pyramids of Egypt in front of your very own eyes. **The experience that you have is priceless, and one which will NEVER be able to be substituted, nor transmitted from or to another source (people, media, etc.) fully via methods such as photographs (though yours will indeed serve as lovely reminders and memories). Because it is your UNIQUE experience.** You will come to truly understand the timeless expression, "pictures just don't do it justice."

Standing in front of these massive triangles of limestone, I was awestruck. 'How was it possible for these structures to have been built…especially during the time that they were, with much less technology than we have today?' I asked myself. My mind took a trip back in history and tried to figure it out. But it couldn't. I listened to dozens of tour guides offering explanations to tourists. "And new artifacts are constantly being discovered," I overheard one guide say with a sense of mystery in her voice, "so perhaps" she continued, "the true secret to how the pyramids were built is written in a document somewhere, waiting to be discovered."

FASCINATING.

Fascinating because I had never thought about that. I had spent a significant amount of time in The Egyptian Museum in Cairo just a few days before visiting the pyramids. That place felt more to me like a warehouse than a museum. There were artifacts and statues on display, but many were on the floor and covered with plastic. New statues, artifacts, and other ancient items were still constantly being discovered during excavations. Therefore, what that tour guide had said was totally plausible…

Actually, I was unable to leave Cairo having seen the Giza Pyramid Complex just once. My curiosity had grown. I decided to go twice. I was so taken aback by these structures that I figured I should give myself time to approach the learning from two angles -- from a guided tour angle and from a personal exploration angle as well.

The Bent Pyramid had recently opened up, and I had the opportunity to go inside. 'Could I find riches inside?!' I thought. I would INDEED find riches…and they would be RICHER THAN I EVER COULD HAVE IMAGINED…

There were no riches in sight just yet, but I was persistent, and I knew I had to keep searching to find them. Just being afforded the opportunity to climb through narrow passageways to reach chambers where riches were once stored was a marvelous experience. Re-living history was breathtaking (literally, due to the physical strength required to maneuver around inside the narrow passageways of the pyramid). I woke up incredibly sore the next day from all the crouching I had done the day before to get around inside the pyramid. My legs were in excruciating pain because I had not used those muscles in those ways for the majority of the trip, yet I was content. The experience was totally worth it.

Let's talk about the shape of a pyramid. It is a majestic shape. One of symmetry. One of structure. One of unity. One of power. When I was shown the papyrus plant during the tour, I learned that the end of the plant has a pyramid shape to it. 'Hmm...' I thought to myself, 'so this shape is actually derived from nature...I wonder if and how natural forces may have influenced the shape of these amazing structures...'

And then, I found them. FINALLY! The riches I had been searching for.

Hieroglyphics. Wow! I was not only awed by the beauty and color of the drawings on the walls but also by the details involved. But I received far more than just a visual experience. My guide actually translated the hieroglyphics into English for me so that I was able to understand what the ancient Egyptians were communicating through these writings oh so many years ago. I came to understand the ancient Egyptian counting system, how humans interacted with certain animals, which ones, to what capacity, and why. This was truly incredible.

But actually, the hieroglyphics, while indeed a form of riches, were not the TRUEST form of riches gleaned from the experience. The real riches encountered on this journey were the realizations of how powerful living out history in person can be and how that experience can move and change someone.

What I learned about Egypt and the pyramids, due to having an "outside-of-the-classroom" learning experience, stuck in my mind much better and more clearly than when I had read about it in a textbook in school years prior. THIS I believe, is a MUCH more powerful way of learning. EXPERIENTIAL LEARNING. Where we are actively involved, and participate in the learning process. I took in the smells of Egypt, physically touched the limestone blocks of the pyramids laid down many years ago, experienced the struggle and emotions associated with crawling (or squeezing) through extremely narrow pyramid passageways that many before me had done, and so on. Textbooks, documentaries, etc. are wonderful sources of knowledge however I believe that they should supplement the learning that takes place when one has the experience for oneself (whenever possible).

Let's talk about beautiful beaches. Who doesn't love those? Many folks love to go sit on a beach. Relax. Daydream. Do nothing. And that's fine. I would never have discovered one of the world's most beautiful beaches had I not decided to travel to The Philippines.

Have you heard of Seven Commandos Beach? I would be very surprised if your answer is "yes."

I thought paradise didn't exist in real life. Until I encountered Seven Commandos Beach.

A true paradise in every aspect of the word. '...but could this be? Is it possible that a place like this truly exists?' I wondered. Little did I know, before visiting this amazing place, that such amazing treasures truly DO exist in our world...I just hadn't discovered them yet... But that afternoon, I discovered an even greater treasure than the beach itself...

I landed on Seven Commandos Beach during an island-hopping tour that began in El Nido in The Philippines. During the time I was there, El Nido was just beginning to become a backpacker location. When I arrived, I found a local family looking to host some backpackers, and that is where I met some incredible people -- both Filipinos and foreigners -- amazing souls who I never would have come into contact with had I not ventured out of my comfort zone. But we'll get to talking about the amazing people you'll meet through your travels in just a short while. Let's focus on the location for now.

The vivid image of pure paradise still remains in my mind. Impeccable white sand. Serene blue water. And what a perfectly sunny day it was. A hammock swaying in the wind. Many actually. I chose one. I laid on it. I relaxed. Swaying back and forth in the gentle breeze, watching the waves come into shore and then retreat back out to sea. I later moved out of the shade and into the sun. And sat down in the sand. But not before purchasing a coconut from the only vendor on the island -- a small hut, but a memorable one. The coconut's water was cool and refreshing. Absorbed in the warmth of the sun, sandy and slathered with the usual obscene amount of sunblock, I was content. Content beyond belief. The pristine white sand, the cool blue water, the gently swaying hammock, and the delicious coconut would unfortunately not return to The United States with me however the memory of this seemingly unreal experience would.

But what about the treasure I discovered that was even greater than the beach itself? How could there be a greater treasure than that?

The *true gem* I encountered on that beach was a lesson. The lesson that in order to find true treasures in life, we must get out of our own way and out of our comfort zones.

Another memory of a place I visited that stands out clearly in my mind is Christiania. What is this place? Well, by the name alone, one might think that it is a holiday, a religion, or a combination of the two. Yet this place actually couldn't be further from the concept of anything even remotely related to either of those words.

There is something odd about this place. Very odd. Extremely odd in fact...

I biked through the city of Copenhagen during my Europe Trip and made a stop at this unusual location. When I first entered this zone, something felt different. But I couldn't quite place what it was. As I continued to explore, the feeling of eeriness grew inside of me. I found myself unable to explain this feeling and what I was experiencing.

Filled with more than a few "unique" characters (I'll let you go there on your own and define the word "unique" as you see fit), it is a place that runs autonomously from the government. My understanding is that those people residing inside of this zone follow their own rules, without regard to local or national law.

The theme of being on one's own surfaced. Individualism. Following one's own path. Even if it's hard. Even if it is different than that of other people. Than that of the majority. Than that of other groups. Than that of other societies. Than that of other cities. Than that of other states. Than that of other nations. Than that of other worlds waiting to be discovered. I found that revelation to be astonishing.

Those folks play the game of life by their own rules. Living the way *they* determine their lives should be lived. I realized that *I* do the same. So why can't *you*?

Although it initially appeared that I had ABSOLUTELY NOTHING in common with this town or its peculiar residents...

somehow, I DID. I was able to relate to these people. To their town. To their world. To their ideals, values, and beliefs. To their burning desire to have and maintain independence. I was finally able to place that eerie feeling I had felt upon entering Christiania. Deep down, I had experienced a feeling of connectedness. To a group of people whom I knew almost nothing about. In a place that just several days before I didn't even know existed. Somehow, albeit coming from very different backgrounds...we understood each other...

The pyramids in Egypt, Seven Commandos Beach in The Philippines, and Christiania in Denmark are just a few of the amazing places that have greatly impacted my life. If amazing places such as these can change my life, amazing places such as these can change your life too!

ARE YOU CURIOUS ABOUT PEOPLE?

Oh yes, the people you will meet. I have met some of the most incredible human beings on this planet through travel -- local people within the communities I have visited, as well as other travelers and people working abroad. I have made and maintained lasting friendships that have changed my life and have opened up my mind to new ideas and perspectives.

When traveling throughout China, I made a dear friend. He was very curious about American culture, and I was curious about Chinese culture. Our interactions represented a continued cultural exchange, along with a language exchange. He taught me some Mandarin Chinese, and in return, I taught him some English.

It was a rainy day. Definitely not ideal conditions for a hike. Definitely not ideal conditions for a hike up an enormous mountain. Especially for someone as ill-prepared as myself at the time. Out of shape and without the proper gear, I did what any sensible person would do. I decided to do the hike. To both ascend and descend in a day. A rainy day. A slippery day. A day when nobody should be climbing a mountain of any size.

That day marked one of my greatest accomplishments in life.

We trudged up the mountain, he walking much quicker than I. I had to take many breaks. It was cold and wet, and I was constantly slipping. Struggling to maintain my footing became a consistent issue. But I carried on.

And when we reached Lotus Peak, the summit of Huangshan (Yellow Mountain), we took a picture in front of the elevation sign: 1864.8 meters, which is nearly 6,120 feet.

It was with him that I experienced one of the most challenging physical excursions I have ever embarked on and completed. And while that was indeed a true accomplishment, it is not the one referenced at the beginning of this story.

The real accomplishment here was learning the value of pushing someone to achieve great things and allowing them to push you to do the same. The value in both holding someone else accountable and allowing someone else to hold you accountable for achieving goals. Goals that require surmounting some of the most daring, bold, and significant hurdles -- be they physical, mental, spiritual, etc. -- that you will face and need to overcome throughout the course of your travels and throughout the course of your life.

And the synergy that is created between those parties when that bond is created and nurtured. Irrespective of homogeneity in gender. Irrespective of homogeneity in race. Irrespective of homogeneity in sexual orientation. Irrespective of homogeneity in religion. Irrespective of homogeneity in age. Irrespective of homogeneity in country of origin or residence.

RESPECTIVE of homogeneity in the genuine desire to connect authentically with another human being.

In a part of China called Shangri-La, I found myself wandering around an open field one afternoon. I came across several tents and decided to peer into one of them. And inside this tent, I found something amazing!

I encountered a woman sporting cowboy attire (like most of the others in that place). She had a long red shirt on that looked quite old. It appeared all of her clothing was old. Perhaps her clothing consisted mainly, if not solely, of hand-me-downs.

Her skin was very dark. Her face covered with wrinkles. Deep brown eyes. Her two front teeth a golden shade of yellow. Her hair braided intricately. A necklace of beads around her neck.

She ushered me into the tent and instructed me to sit down. Then she offered me freshly made yak butter. I didn't want to put my tastebuds through this however I also did not want to be rude and refuse her offer. So I tried it. And to my surprise, it was actually quite tasty!

It would have been ill-mannered of me not to offer her anything in return. But I had nothing to give. Let me rephrase that. I had nothing OBVIOUS to give.

So I did something that some people will view as resourceful, and others will view as absolutely disgusting. Noticing that she was barefoot and with nothing else I could think to offer her, I proceeded to remove my socks and gifted them to her. This action caused a large smile to appear on her face.

The woman I encountered in that tent was most certainly wonderful. However, the amazing item referenced earlier in the story that was found in that tent on that day, was a connection. A real connection.

A human connection. A connection between two people who didn't speak the same language. A connection that flourished through the use of gestures and the exchange of gifts, as strange as they were -- salty yak butter and a pair of sweaty, worn socks. Although a strange interaction indeed, it was a fascinating cultural exchange that will forever remain ingrained in my mind. And I am confident that it will forever remain ingrained in hers as well.

Let's drop back in on India. Before going, I dreamed of all the amazing sights and sounds, of yoga and meditation, of amazingly delicious and spicy food. It was a new place though, and I had heard that there were dangerous parts. I was scared of the unknown. I was more scared however that I might not have the chance to visit India for this length of time again, so I opted to go. I'll never forget my cousin's warning that came via email shortly before my trip. "If you aren't prepared, India will eat you alive..."

She couldn't have been more accurate on that one. I encountered many people who used deceitful tactics in an attempt to take advantage of me throughout my time in India. It often seemed as if people were attempting to trick me everywhere I went. It was exhausting. I was in constant negotiations for everything from hiring a tuk-tuk to purchasing a cup of chai. At the time, this was extremely irritating. However, the knowledge I gained and the growth I experienced -- particularly in reading people and pertaining to haggling and negotiation -- was extraordinarily valuable and only noticeable upon reflecting on the experience afterward.

I got to India and EVERYTHING went wrong. After leaving the airport's main building, I realized I needed to go back inside to pick up some information however I was stopped by police officers.

I was told firmly that I could not re-enter the building with luggage. Fortunately, I found someone who had been on my flight, and she agreed to wait outside with my belongings while I acquired the necessary information inside.

After hours of riding around in a tuk-tuk trying to find a place to sleep for the night, I was brought to "The Official Tourist Office" which was in no way, shape, or form official. I was told that there was nowhere to sleep that night since a festival called Diwali was taking place and had therefore sapped up all available hotel rooms. I was however offered a "special deal" wherein I could stay one night in a hotel in Delhi and then be off to Kashmir the next day for the experience of staying on a houseboat. This hotel where I would stay (in Delhi) for the night, the gentleman in the office claimed, was not currently open to the public, however he could make an exception and open it up for me to stay the night if I agreed to purchase the package deal... Although the excursion seemed extremely overpriced, it was way past midnight, and I was unaware of any other options. And I wanted to get some sleep. So, I reluctantly agreed to the package deal, including a place to stay for the night in Delhi.

At the time, I believed what I was told about the hotel occupancy situation but would later find out it was not true. I would get taken to Kashmir the following day. Unaware of some pretty serious details. And had I known about those details prior to making my decision, I would have absolutely chosen NOT to go...

I arrived at the houseboat in Kashmir and met the family who owned it. I also met a Japanese tourist on that houseboat who changed the course of my life. He introduced me to his friend, who was living in Tokyo at the time. That connection led to me creating a strong friendship, making the decision to relocate to Japan after the conclusion of my trip, and to finding my first place to stay in Tokyo.

The father of the family on the houseboat appeared nice at first, but after a few days of staying there, I began to feel uncomfortable. Something seemed off about him...

I was offered an opportunity to go on an overnight hiking trip in the Himalayan Mountains with the father of this family. Something made me feel uneasy about taking a trip with this guy, but I let the thought of being immersed in such beautiful nature outweigh my internal sense that something wasn't right. Reluctantly, I agreed to go. 'What could really go wrong?' I thought to myself...

We started our journey toward the mountains and were driving along dirt roads for a while. We then stopped at a local market to pick up some food, and the atmosphere began to change. People's faces were now completely covered; people's movements were erratic and aggressive; an aura of tension was lingering in the air. As we ascended into the mountains, we began to pass checkpoints manned by armed guards, and my anxiety level skyrocketed. My heart began beating faster and faster, feeling as though it would burst out of my chest. I started to feel that I could be on the verge of being taken hostage, or worse off, eradicated. Scary thoughts began to race through my mind. 'Why is the father of this family so secretive? Is he collaborating with someone else, or with some organization? What if it is my American passport he is after? He mentioned he doesn't like Americans...why? And how strongly does he dislike Americans? What will he do about his dislike for Americans, and could it be taken out on me somehow? Will I be captured? Will I be tortured? Will I be killed?'

Something felt wrong. Very wrong. I knew I had to leave.

When we arrived at the spot in the mountains where we would stay, I told the father of the family that I wanted to go back to the houseboat. He detested the idea and tried to convince me otherwise, however once he sensed that my mind was made up, he instructed our driver to take me back.

The driver, someone who worked with the father of the family on that houseboat where I had been staying, spoke no English. And we had one of the strangest rides of my entire life back to the houseboat. Without warning, he pulled off to the side of the road in front of a house, got out of the vehicle, and motioned for me to do the same and to follow him. We entered a house inhabited by gypsies in the mountains, and within just a few moments, he had disappeared upstairs with two of the gypsy women, leaving me by myself to interact with a woman (perhaps the mother of those two women) and a small child.

I did my best to communicate with her. It wasn't easy, but I managed to do what I could to overcome the language barrier, using gestures to get my points across. Here I was, this strange man in her house. Here she was, this strange woman in front of me. Holding a child. In front of us, a blazing fire and way too much smoke to be considered acceptable by any standards. Two people in the same room. From different cultures. From entirely different upbringings. Yet somehow, there was an understanding between the two parties.

When the driver came back downstairs, which felt like hours later, he left the house. And I followed. And we continued the journey back to the houseboat. Back to the houseboat we finally arrived, and off to the bus stop I would go the following day to leave Kashmir.

There are two important facts that I learned shortly after this experience. I believe they are both relevant and important for you to know as well:

1) Kashmir is a disputed territory. One that has experienced serious conflict and has been historically prone to major unrest and violence.

2) The spot in the mountains we were in, where I felt so uncomfortable, was a known terrorist hideout zone. It was just one mile away from Islamabad in Pakistan, where United States Special Forces had tactfully orchestrated the elimination of one of the most infamous and influential terrorists of all time. *I had been just one mile away from where Osama Bin Laden had been wiped off the face of the Earth a few months prior.*

From this experience, I learned to pay greater attention to my surroundings. I learned to scan my environment for threats. And I learned to be more in tune with and trusting of my instincts.

Fortunately, there is a happy ending to this story! I would go on to meet a great guy from Washington State at the bus station – he was on his way out of Kashmir too! While he and I had our differences, he was a great travel partner, and we traveled India together for quite some time. Through him, I would go on to learn just how helpful, valuable, and arguably necessary having social support is in life. Especially during difficult situations.

Along the ghats in Varanasi, India, we came across a part of town where there were birds in cages. From watching and listening, we surmised that what was happening was that the ability to free these birds was being granted (mainly to tourists, but to anyone interested in paying, really). 'Why were these birds caged up in the first place?' I thought...

My new travel partner walked up to the boy holding a few of the cages and inquired about them. About the types of birds being held captive. About what the reason behind it was. I guess my friend was curious too... Not only was he curious, but he was determined to take action. And he was determined not to pay for it. What my travel companion did next would shock everyone...

After action was taken, off we walked past the boy and the market. But the boy followed. He coaxed us. He picked up a stick and threatened us. He pleaded with us, claiming that his master would kill him if he returned empty-handed. When that didn't work, he began throwing rocks at us. And following that, the pleading turned into crying. Begging. A local took notice of the situation and began to follow us too. Another local saw what was happening and began to follow us as well. And another. And then another. And before we knew it, an angry and agitated mob had amassed and had begun following us.

Finally, my partner stopped walking and agreed to pay. Not the full fee. But half. But both parties were satisfied enough. The boy would be able to return to his master with something, and my friend and I would live to see another day.

It wasn't the fact that my travel partner had freed the owl that upset that little boy. It was the fact that my friend had refused to pay for doing so.

In my new friend's mind, he was a hero. He had just freed an enslaved creature. Coming from his background working with animals, he felt that what he did was justified. That it was right. That NOBODY should have to pay to free ANY animal. That animals should be free in the first place. Just as they are when they are born into nature.

To this little boy, this was his livelihood. He was trying to make a living for himself. To earn enough to eat and survive. To return to his boss both without the bird and without rupees (the local currency in India) surely would not have gone over well. He would likely have been reprimanded. Perhaps beaten. And as the boy had said, his life could have even come to an untimely end. Perhaps this boy didn't feel he had any other work alternatives available.

Perhaps he didn't see anything wrong with the work he was doing. By providing people with the opportunity to free an animal, perhaps he felt he was doing good in this world. He was allowing justice to be served. HE was the hero.

From that situation, I learned how to handle stressful situations better. I learned that panicking isn't effective and that doing so only freezes our reasoning faculties. I also learned how to better assess situations from multiple viewpoints and to try to view them as objectively as possible. To aim to see both (or more) sides of the story. And to work together collaboratively to come up with a solution that is desirable or at least agreeable to all parties involved.

There was another memorable incident I recall from traveling with this character. And this incident pertained not to the freedom of a bird. But to the freedom of a human.

We were ready to hop on a bus to our next destination. But before doing so, there was just one minor thing my new friend from Washington felt he needed to do. And that was to steal the hat that was hanging on one of the handlebars of a police motorcycle. Not a big deal, right? Things get stolen all the time. It's just a hat. And the officer probably wouldn't even notice. And even if he did, he probably wouldn't care...right?

Wrong. The police officer took notice. And the police officer took great offense. He grabbed my new buddy and began angrily shouting at him in Hindi (the local language).

A crowd gathered to investigate the ordeal, and we once again found ourselves surrounded by a group of people. But this time it wasn't the owl's freedom that was at stake. It was my friend's.

One man who spoke English well told me that the officer was very seriously considering taking my friend to jail. He said that if we gave him a certain amount of money however, that he would give it to the police officer and arrange to have my friend let go.

This was a difficult situation. Forking up a rather exorbitant (by Indian standards) sum of money for this, or in essence, bribery, is against my morals. However, refusing to pay and allowing my friend to go to jail (and missing our bus to the next city) didn't seem like a phenomenal idea. Although I felt uncomfortable with the decision, padding someone else's pockets in order to preserve my friend's freedom seemed to be the best choice available.

I checked my pockets. Nothing. I asked my friend if he had any rupees on him. Nope. He was completely out. CRAP! 'What are we going to do now?!' I thought. I guess my friend would have to go to jail after all…

And just before I gave up hope, I remembered that I had been keeping a reserve supply of cash on me just in case an emergency like this happened. In my left sock. Strange. I know. But if you attempt to rob someone (please don't), are you going to look there? Point proven.

So I paid the requested sum of money to the man who spoke English and who had promised to diffuse this altercation. The officer grudgingly let go of my pal. Onto the bus we walked. And off to the next city we went.

Here, again, we find a fascinating example of perspective. Instead of just two perspectives in this example, we find three.

To my friend, what a great opportunity to grab a souvenir. And what an amazing story to accompany it! His friends back home would never believe how brave and daring he was! How authentic this souvenir was. The value assigned to it would be so high because it wasn't bought from a store; it was taken in a real-life situation. In my friend's mind, the officer wouldn't find out about this. And even if he did, he wouldn't mind because it would be easy to replace a hat. They are lost and stolen all the time, so this would be a commonplace occurrence.

But not to the police officer. To him, this was a high insult. An unwarranted theft of property. State property. Government property. But most importantly, HIS property. And along with his property came his pride. How DARE this American insult him by stealing a vital piece of his uniform. To him, this was in essence a theft of a part of who he was. And surely he would not tolerate that behavior. There would indeed be consequences in store for anyone looking to disrespect his state and country, let alone him as a person. There would be a price to pay…a BIG price to pay…

And how about the ingenious mediator who stepped in? This ordinary citizen who saw an opportunity. Who saw this situation evolving and realized there was a conflict between two parties. And that these two parties were unable to effectively communicate to resolve it amongst themselves. And that, with the ability to speak both English and Hindi, he could facilitate communication between both parties. As an intermediary, he could take the time to understand both sides of the story and could provide a resolution to the dispute. And in relaying the price required to settle this contentious matter, he could build in *his* fee. After all, why would he NOT deserve to be compensated for his effort?

This incident allowed me another chance to view an interesting situation from multiple angles. It furthered my ability to stay calm in tense situations. And to see the value in making decisions and taking actions that allow for the resolution of conflict in the most peaceful of ways possible.

In Goa, India, my new pal rented a motor scooter and asked if I wanted to use it to check out the town. I didn't. I had never driven a motor scooter before and knew it was extremely dangerous to do so. I also knew that hordes of people were walking around the town with casts and burns from accidents involving these scooters. So, I did what any logical person would have done.

I said "yes."

My inexperienced driving became evident immediately as I swerved through the streets with a complete lack of control. Veering haphazardly from the left side of the road to the right. Back to the left and back to the right. I should have sought out a class to learn how to drive the thing, or at least have insisted that my buddy show me the basics before heading out on the town.

And as I was going along, I noticed an Indian woman selling rugs on the street. Due to my inability to control the vehicle well (practically at all), our first (and final) interaction was not a friendly one.

Because I ran her rugs over. Many of them. Not because this is a hobby of mine (is that anyone's hobby?). Or because I enjoy destroying people's property. But because I could not control the vehicle properly.

Naturally, she became enraged. She began yelling in a language I couldn't understand. I was scared. In the heat of the moment and perhaps as a protective mechanism, my body chose flight over fight, and I fled the scene. But I could have stayed without fighting. I could have been a caring and considerate young fellow and addressed the situation I had created. Not only *could* I have stayed, I *should* have stayed. I should have helped. I should have taken responsibility for my actions. I should have at least paid her for the damage done. Just moments later, I would pay dearly for the lack of caution and concern I had shown for others in that situation.

After speeding out of that area, I came upon a narrow bridge. With a car on it. Headed my way. I got stuck in a pothole, and in an attempt to rev out of it, I slammed into the oncoming vehicle. The impact wasn't overly forceful, but it was enough to knock me off the bike. I was shocked at first. I stood up. I examined the bike, which seemed to be fine. I examined myself. I seemed to be fine too.

And then I noticed my right index finger suddenly resembled a french fry. With ketchup on it. And couldn't be bent past a certain angle.

Had I just stopped to help the woman with the rugs and pay for the damage I had caused, perhaps karma would not have rewarded me in the way in which it did.

I wasn't far from where I was staying, but driving with a broken finger, combined with a horrendous ineptitude to navigate the streets, was not a winning combination. Luckily, my friend was around when I returned. I explained what had happened. He hopped onto the scooter to drive. I hopped on behind him. And off to my first Indian hospital I went. Praying that we would not get into a second accident on our way there.

When I think of a hospital, I think of a clean and sanitary place. Where people who are ill are being treated and helped. What I found there was not that. AT ALL.

Several people were leaving the hospital as we were arriving. Some with rainbow looking burns on their legs. And I'm not exaggerating. A full-on rainbow had been created from the skin that was removed and/or distorted from being burned by the exhaust pipes of these scooters.

My friend and I remained in the waiting area for a while, and eventually, out came a nurse who took us into a room. When we entered, there was an enormous pool of blood on the floor. We were quickly escorted out of that room. This was not turning out to be a very positive experience so far...

And then we were brought into another room. After waiting for quite a bit of time, a doctor came in. He examined my finger and proposed two options:

1) Align my broken (index) finger with my middle finger and bind them together so that the bone could heal naturally

2) Have surgery

The doctor recommended avoiding surgery if possible because it entailed the risk of infection. And with the humidity present in this environment, an infection could be very dangerous as it would carry with it the potential to spread quickly and become difficult to control.

His logic made sense to me, so I decided to avoid surgery. What I didn't quite understand was why, after my two fingers were bound together, half of my arm (up to my elbow) would need to be put into a cast. What I did realize however, was that once I stopped trying to make sense out of things in India, things just worked out better -- I was better off when I didn't ask "why?"

I had come to a crossroads. I could continue my journey or cut my trip short and return home. I no longer had access to my dominant hand. That meant there would be no easy way of writing (emails, letters, signing documents, etc.). If I were to continue my journey, I would have to find a (seemingly impossible) way to carry my luggage from place to place. To make things worse and even more embarrassing, I would need to shower with a bag over my arm because the cast couldn't get wet.

It would make a lot more sense to return home and let my injury heal. I was scared of continuing to travel in that kind of condition anyway. I was worried about the challenges I would face and was concerned that I would be unable to surmount them. The right choice seemed clear.

Fear had indeed accomplished its goal. It had motivated me NOT to give up.

So I continued my journey, pushing on with a ridiculous and seemingly unnecessary cast on my right arm. I took longer than normal to write emails, pecking away at keyboards with my left hand. But I became better at it over time! I showered every day with what became my new fashion statement -- my epic plastic bag. And I found new ways to carry my backpack from place to place, sometimes even with the help of others! I did this for four weeks in India and for an additional two in Thailand before the cast was finally removed.

From my interaction with the woman selling rugs, I learned the value of being a more caring, cautious, and thoughtful person. I recognized the need to stop and help others. To show compassion. To be responsible. To admit mistakes. To apologize when wrong. To work to make things right. And to better assess which risks to take and which not to.

From being injured, I learned that I could find new and creative ways to accomplish things. I could increase my ability to write with my left hand through practice. I could find novel, creative, and effective ways to transport my items from place to place. I could achieve seemingly unachievable things. I could CHOOSE to see myself as handicapped, or I could CHOOSE to see myself as empowered. And I would act accordingly. And I would also be treated accordingly.

From this situation, I became intimately aware of the overwhelming power of fear. We can let it immobilize us, or we can allow it to motivate us to overcome the (sometimes seemingly impossible) challenges that life throws our way. We can give up in the face of problems, or we can keep on going. We can throw in the towel, or we can continue to fight. Whether we decide we *can't* do it, or we decide we *can* do it, we'll be right. Don't allow fear to freeze you in your tracks. Instead, allow fear to push you to accomplish greater things in life than you EVER imagined possible. If I can do it, so can you.

While my new friend's mischievous actions created several stressful situations, he helped me in many ways. He helped me get around with my bag while I was injured. He helped keep my spirits up as we constantly faced the challenges of bargaining and dealing with the dynamics of the foreign culture we were interacting with. He provided a support system. Friendship. And endless laughs. He kept me sane (and I kept him sane too). For these things, I will be forever grateful.

Both my travel partner and I became gravely ill at the same time. He helped take care of me, and I him. We bonded. We became brothers. When it felt like one could no longer go on, the other provided the encouragement and support needed to power through. And vice versa. Through my experience with him, I learned that going through an extremely difficult and challenging situation with another person allows for a special type of strong bonding to occur. One that can bring strangers very close together. And in a surprisingly short period of time.

I met another influential character in Goa, India. Christian Wenge is one of the funniest people I have ever met and had the pleasure of getting to know. There I was, with my entire arm wrapped up in a cast due to my broken finger. And there he was, nonjudgmental, caring, and concerned about what had happened to me.

I had also managed to cut my toe on a piece of glass on the beach. Christian came up with the ingenious idea of using a rubber band to secure a napkin around my toe, acting as a bandage that protected the cut. And it worked! This not only provided a creative solution to the problem but also provided endless entertainment and laughs. As battered as my body was, stuck with a cast on my arm and a napkin rubber-banded around my toe, Christian kept my spirits high. Soaring to new heights, in fact. I wouldn't have wanted to celebrate New Year's Eve on a beach -- carefully trying to avoid everyone due to my absurd injuries and unique bodily decorations -- with anyone else in the world.

And it would be with Christian who I would have the great pleasure of experiencing one of the world's most epic and well-known festivals. He was living in Munich at the time, and years later during my world trip, I found myself celebrating with him again. Yet this time, we weren't ringing in the year 2012 together. This time, we were out celebrating Oktoberfest 2019 together.

And oh how good it felt to be reunited once again with someone I had bonded with through travel. Someone who had gone through the same trials and tribulations that I had. Someone who understood me. Someone who I could trust. Someone who I could genuinely call a friend.

India was by far the most challenging place I have ever traveled to, but looking back, it was also one of the most character-building experiences of my life. At first, India beat me to a pulp. But by the end of my time there, I had grown immensely as a person. I learned to read people and situations instantaneously and to respond in thoughtful and calculated ways. I learned how to innovate and negotiate to solve problems. I learned how to be more aware of my surroundings and enhanced my ability to detect danger and follow my instincts.

I have been fortunate to connect with so many amazing souls on my travels and to create and nurture deep and fulfilling relationships. If I can do this, so can you.

ARE YOU CURIOUS ABOUT CULTURES?

I find culture to be fascinating, and perhaps you do (or will by the end of this book) too!

I had a cultural experience during my time in Maasai Mara, Kenya that I found to be particularly interesting. Perhaps you will find it particularly interesting too.

While there, I embarked on a safari and the experience was unforgettable. The plethora and diversity of animals spotted, the breathtaking sunrises and sunsets, the smell of nature in the air, and the list goes on. On the third and final day, I was introduced to The Maasai People. And I would never be the same again.

"If you can jump very high, you only have to give up five cows. If you can't jump that high, you will owe ten." I recall being told. This statement was in reference to what a man would need to give up in exchange for his future partner's hand in marriage. I was taken aback by these words.

Dressed in traditional tribal garb. Variations of the color red and with different patterns. Some solid, some checkered, and others striped. There they stood. In front of me. Around me. With me. I watched as, one by one, each participant launched his body into the air, touching back down to earth only to repeat the action. Again. And again. And yet again, as if propelled by an invisible springboard.

And I was beckoned to join! I couldn't turn down the chance to participate in this age-old and legendary tradition. Ecstatic, I accepted the invitation!

Adumu is part of Eunoto. Eunoto is a ceremony during which young men become warriors and transition into adulthood, allowing them to marry. How high they jump affects their level of attractiveness to potential future mates.

I didn't come anywhere close to jumping as high as even the lowest jumper. I elevated myself as far off the ground as possible, trying my hardest to retain the level of stamina they seemed to all have, yet found the task arduous. Strenuous. Difficult. Tiring. But alas, it didn't matter. Because I had been afforded the rare opportunity to participate in a crucial rite of passage. I was invited into their world. I became, even if for only a fleeting moment in time, a Maasai Warrior.

After participating in the ritual dance with the Maasai, I was invited to watch how fire was created without matches, kerosene, or any other unnatural items. Through a process of wiggling a stick back and forth in a certain manner on another piece of wood, a tiny smolder awakened, and when introduced gently and carefully to a dry pile of leaves, a flame was born. This was truly an incredible process to watch, especially for someone who can barely light a match. Aside from finding the process of creating the flame and the flame itself to be impressive, I also found the fact that this tradition had been preserved for so many years to be impressive.

And into a Maasai Warrior's home I was invited. And into a Maasai Warrior's home I went. There, I met one of his babies, along with one of his wives. Polygamy (the practice of having more than one partner) is widely practiced among the Maasai people. And in many other locations around the world as well.

While the Maasai Warrior's wife boiled water, the smoke from the flame heating the pot began to get more intense. It started producing an unfathomable amount of smoke. My eyes began to release droplets. Then streams. And finally fountains. As I sipped tea from the mug that I had been handed, the atmosphere was almost intolerable. And then the child began to cry. But through the hazy and seemingly impenetrable smoke and over the baby's cries, I learned of this Maasai Warrior's life, struggles, and accomplishments. And it was fascinating.

This is yet another example of humans from very different walks of life coming together to share tradition, culture, and values with one another, with the aim of communicating and fostering an understanding. It was achieved. It was potent. It was unforgettable.

During my world trip and after visiting Kenya, I visited Uganda. It was there that I was introduced to the Pygmy people. And oh my, what an experience that was.

We rowed past a small island, and when we arrived at the location where the Pygmy people were, I was greeted by a young boy and the daunting elevation of the hill ahead. It was so steep that this little boy, who navigated the angle with astonishing ease, had to hold my hand while ascending in order for me to complete this seemingly impossible journey. This little boy helped me push my physical limits. If he could do it, I could too.

When I arrived up top, I was greeted by the Pygmy people. They did a ritual welcome song and dance for me. The teacher of the village showed me their school. The classrooms were small. He was the only teacher and taught all of the subjects. He was one of the few who spoke fluent English. I didn't have too much money left after donating a fair amount of it to the community after the welcome dance, however I gave what I had left to the school.

I was introduced to the class and afforded the privilege to speak to these children. Eagerly awaiting my words in silence, and with looks of great curiosity, I had their full attention. I did my best to instill the values of education and self-confidence in them within the mere minutes I had. "Study hard" I said, "and you can do or be anything you want in life." As the children's eyes met mine, I felt an intense feeling of hope and inspiration transcend the room. Someone had just changed their lives. Someone had just told them something that they would treasure forever. Someone had just expressed their belief in these children. And in that very moment, they realized that THEY TOO could choose to believe in themselves. And if they could believe, then they could achieve. If they can do it, so can you.

I spent some additional time with the children after that speech. They showed me around their school and classrooms. They were proud. Curious. Knowledge seekers. Educators. I was just as much their student as they were mine that day.

Eventually, my tour guide mentioned that it was time to leave, and as I received goodbyes from the community, that same boy who had helped me up the hill appeared in front of me. He stared up into my eyes, and although we were unable to communicate in English, we were able to communicate with our souls. And from his gaze, my heart heard him say, "It's time to go now. But this isn't goodbye forever. Don't be sad. Come, let me guide you safely back down the hill."

He led me back down the hill. My hand in his. His clasp tight. He didn't know anything about me. And I didn't know anything about him. But he trusted me with his life. And I trusted him with mine. This boy's helpful demeanor is sure to get him far in life. He is a role model to his community. I was humbled and will be forever appreciative of having met such an inspirational leader. This experience taught me that education is a two-way streak between the "teacher" and the "student/s." That leaders help, inspire, and uplift others. And that leadership doesn't have age limits.

I had another amazing cultural experience with a tribe in the Amazon rainforest in Peru. They wore hay skirts. And hay tops covered their chests. Their hats and headbands were made with hay too. Multicolored feathers stood staunchly in the air, adorning the head accessories of the men.

I was afforded the opportunity to join their ritual dance. Holding hands, we danced around in a circle, bringing our hands up and then down as we sang in the muddied arena of earth that an intense rainstorm had just created. Fortunately, during the storm I had been given an enormous banana leaf. This leaf was so large and sturdy that it acted as an umbrella, shielding my entire body from the onslaught of nature's water.

And then I was handed an apparatus from which I was told to shoot a poison dart. With no actual poison on this dart. My target? A tree. Success? Indeed! These darts are traditionally rubbed against the skin of poison dart frogs with the aim of delivering a lethal dose of venom to the target upon impact.

This cultural experience showed me the pride, openness, and authenticity this tribe has in sharing their culture with others. They welcomed me in with open arms, eager to share their values and rituals. They showed me how they lived, their hunting practices, and so on. Through these rituals, I was able to understand the importance placed on connections among human beings and connections between humans and the environment; it was clear that there was an emphasis on maintaining harmony in relationships. I left feeling that I had encountered a brave group of warriors, not only in the physical sense but in the emotional and spiritual senses as well. From this experience, I was inspired to be more open about myself and vulnerable with others and to share my rituals and experiences more freely.

Trudging through the Amazon Rainforest was tiring. It was scary. Almost everything we encountered could have killed us. My days there provided me with an understanding of a new way of life. I learned how people in that environment survive and thrive. I learned the many ways in which the tribes derive resources such as food and medication from their environment. For example, some medicine is extracted through specific trees. Fascinating. Through this experience, I became a more resourceful person.

I had yet another interesting cultural interaction with the Long Neck Karen tribe in Thailand, who originally emigrated from Burma. I observed bright and vibrant clothing, with beautifully colored cloths tied around their hair. The most eye-catching detail however, was the length of their necks. They were long. MUCH longer than I had ever seen before. I tried not to stare. I wished not to be rude.

These ladies had varying amounts of metal coils around their necks, which increased in accordance with the age of the girl/ woman. Far different than a necklace, brass coils were present. And these rings were piled one on top of another, from the bottom of the neck to under the chin.

The people in this tribe use these rings to increase the length of their necks. Or so it appears. In actuality, the length of the neck does not get altered. Instead, the frame of the body gets pushed downward due to the pressure caused by the rings, giving the appearance of a longer neck.

These coils are a signature mark of a tribeswoman. The girls begin to use these rings when they are young. They hurt, but pain is a requirement of belonging to the community. The reason why the tradition began is unclear. It has been speculated that the neck coils were used to prevent slavery by making the women less attractive to other tribes, to distinguish the tribe's cultural identity, to prevent bites from tigers, amongst several other theories. This tradition was once the cause of persecution. Today, it represents the tribe's commitment to maintaining its traditions. This seemingly bizarre tradition twisted my mind in directions it had never been before and taught me to keep an open mind about other cultures and traditions.

And so while I was unable to communicate in the same mother tongue with these lovely ladies, I still exchanged gestures and looked into their eyes. And they looked into mine. I could somehow understand something. Something about human nature. Something about love. About loss. About hardship. About tradition. Something hard to put into words. Something you need to experience yourself to fully understand. If I can experience this, so can you.

Some of the most fascinating cultures are not solely found in fairytales. They are real. They are out there and they are waiting to be discovered by you. If I was able to discover them, so can you.

ARE YOU CURIOUS ABOUT NEW EXPERIENCES?

I'm going to break experiences down into three categories -- encounters with wildlife, encounters with food, and encounters with adventure sports.

WILDLIFE

I have been fortunate to have had many interesting and engaging experiences with animals in different parts of the world.

Communicating with dolphins in the Bahamas was an incredibly rewarding experience. I had heard how intelligent these creatures were, but it wasn't until I interacted with them personally that I realized just how truly incredible they are. The ability they have to learn, understand, and respond to the commands that their trainer gives is unreal to watch. They communicate with sounds -- whistles and clicks -- and their system of communication is both complex and impressive. They are sweet and friendly to be around. Their intelligence, warmth, and sheer beauty is astounding.

Being on the great plains of the Maasai Mara in Kenya was yet another incredible opportunity to interact with wildlife. I got within feet of a lion. I observed its ferocious and deadly teeth. I stared straight into the eyes of one of nature's fiercest predators. This wasn't National Geographic. This was real life. This was MY life. This was MY story.

I was moved when I learned that lions can actually identify someone from the Maasai Tribe with their keen sense of smell. And when they smell a Maasai in the wild, they do not attack. Because the lions respect the Maasai. And the Maasai respect the lions. This is a powerful example of the strength of an interspecies bond and understanding.

I also observed two cheetahs lapping up water together at a small pond. I spotted a leopard in the trees laying lazily across a branch to catch a respite from the scorching sun. I observed giraffes leisurely crossing the road, taking their sweet time -- I was in their territory; I was in their world.

And the above paragraphs highlight just a few of the amazing encounters I had with African wildlife.

My half-day with the elephants in Phuket, Thailand was unforgettable. All of the elephants in the sanctuary have been rescued from various locations, and although the fee to spend a day with them is not cheap, it goes toward rescuing the elephants, maintaining the facility, providing the staff with a livelihood, and various other things I feel are important.

Participants get to feed the elephants, help give them a mud bath, jump in the pond with them to help them cool off, and finally, get into an extra-large outdoor shower with them to clean them and rinse them off. Having the opportunity to spend time with these gentle, intelligent, and perceptive giants was such a privilege.

I learned that each elephant has a mahout or human trainer/caretaker. Each mahout is assigned to one elephant and they care for that elephant for life – until either they or the elephant transition from the physical world into the spiritual world. This is a serious responsibility, and therefore the turnover rate for these jobs is extremely low. As a matter of fact, the mahout for an elephant usually comes from a family whose members have been doing this job for generations. This was a big "wow" moment for me.

The elephants all have their open pens (rooms) where they sleep at night. During the day, they are free to roam the forest. They are not shackled or chained in any way, shape, or form at any point in time. I found this to be such a humane practice and wondered if it could/should be applied to all animals and in all settings.

I learned that tourism involving riding elephants is quite harmful to them. Since learning this, I have begun researching the activities I want to do involving animals and making ethical and informed decisions on how I spend my money while traveling.

Gorilla trekking in Bwindi Impenetrable Forest in Uganda. Another activity I was unsure I would do. The journey from Kampala to the forest was beyond inconvenient. So much so that I almost didn't go. But I did. And I am so glad that I did. I decided to hire a private driver. This turned out to be a great decision, and I would end up making a lifelong friend out of this experience as well.

Waking up before the crack of dawn wasn't particularly fun. Neither was driving on the insanely bumpy roads leading to the forest. But interacting with my driver was another interesting and amazing experience. We exchanged stories and ideas, and I came away from the experience with yet another new friend. This type of cross-cultural exchange is powerful. And the ability to assess situations objectively, to the best of our ability, in order to be fully receptive to new ideas, crucial.

A group of gorilla trackers (humans) went out to search for a family of gorillas. Our group was comprised of tourists, a guide, porters, and a guard. The guard was armed with a gun to shoot into the air and scare off wild elephants if we were to encounter any. This is because wild elephants are known to be aggressive and charge. We were told that the hike could be easy or intense. Short or long. Depending on what part of the forest the gorillas were in and what the weather conditions were like the day of the hike. Fortunately for us, the trackers spotted a family of gorillas within a relatively short amount of time. The weather conditions were decent, and the gorilla family was located in a spot that was not extremely taxing on the body to get to. We walked. And walked. And walked. And then we found them. And when we did, it was a truly amazing and memorable moment in my life.

Coming face-to-face with these amazing primates in their natural habitat was a breathtaking experience. I got very close to them. It was strange. It was different. It was wild. After all, we were in the wild. I was in their territory.

Upon first sight, they were in the trees. An abundance of flatulence. Massive amounts of defecation. I was bewildered. They were proud. They knew we were there, and they didn't care. Which is why we were able to get rather close, though instructed not to get too close so as not to frighten them.

At Bwindi Impenetrable Forest, tourists are only allowed near gorilla families that have been interacting consistently with humans (initially the park staff only) for quite some time, and are therefore accustomed to having these types of interactions on a regular basis. And the groups of humans allowed to interact with the gorillas are rather small so as not to overwhelm these magnificent creatures. Encountering gorilla families that haven't had frequent exposure to humans could prove to be dangerous. Their reactions are a lot less predictable. And this is why tourists are not introduced to these groups.

Watching the gorillas eat, take care of their young, and interact with one another was quite the spectacle. I thought about how similar I am to them. We are of a different species. But yet, we are not that different from each other at all. As I stared into one of the gorilla's eyes, we seemed to understand each other. Somehow. Some way.

If I can have enriching experiences with wildlife, so can you.

FOOD

How about trying new foods?

You've probably seen tons of cool recipes in some great cookbooks. Perhaps you've watched the cooking channel and have made what they've made. All from the comfort of your own home. Time to try something new. Time to experience food in the place of its origin, with local people and using local ingredients. That's a very different kind of experience and an incredibly rewarding one to have.

I always enjoy trying the local foods when I travel.

I spoke earlier about Egypt, however I left out an amazing dish I discovered called Koshari (also spelled "Koshary"). This is actually Egypt's national dish. It is a mixture of macaroni noodles, rice, and lentils, topped with Egyptian style tomato sauce, fried onions, and garbanzo beans. And I had the pleasure of having it for the first time with an Egyptian. The gal working at the hostel I stayed at was so excited for me to try it, and when I did, I absolutely loved it! Along with a deep yearning to see and learn more about Egypt's amazing history, the food is another reason I will be heading back!

In India, I discovered and subsequently fell in love with and later proposed to (okay, I didn't go that far) The McSpicy Paneer Wrap. Paneer is a type of Indian cheese, and it is put into the wrap along with other delicious ingredients. The wrap is made spicy, in true Indian fashion, and is a delightful experience for the taste buds. I highly recommend it. I have visited many a McDonald's around the world. I would attribute a large portion of this brand's success globally to its ability to understand the culture and preferences of the local market, and to adapt its menu accordingly.

I would be remiss if I did not mention Italy in the food section. Ah, Italy. Napoli to be exact. It was literally an eat-a-thon of the best pizza and pasta in the world during my few days there.

Let's talk pasta first. The rigatoni-style pasta I ordered was just as delicious as you would expect pasta in Italy to be. Fresh. Al dente, and with tomato sauce. But I noticed that the tomato sauce wasn't sauce. At all. It was just smashed tomatoes. 'That's it?' I thought to myself. I had become so accustomed to eating tomato sauce in

The United States that I had expected it to be the same here. But it wasn't. And I was *glad* it wasn't. It was better. No preservatives. Nothing else added. Nothing else needed. Just smashed tomatoes. And it was one of the best meals I've ever had in my life.

The pizzas I sampled came from doing extensive research and finding the top pizzerias in Napoli. The pizza-making process (which I documented via video) is accomplished in just a few minutes. Stretch the dough out, pour on the sauce, sprinkle on the cheese, drop on the fresh basil, and then pour on the olive oil. Into the oven it goes, and out of the oven it comes. Fresh, hot, and ready to eat!

My perception of the key to why Italian food in Italy is so amazing?

1) It's their country so you better believe they know what they're doing

2) They use fresh, high-quality, local ingredients

One of the biggest takeaways I had after eating in Napoli was how much tastier and healthier food can be when it comes from local sources. When this occurs, the ingredients are fresh, and therefore do not contain any preservatives as they do not need to last a long time before being consumed. But no need to take my word for it, just take a look at how many 'Farm to Table' type restaurants have begun to emerge right here in The United States. The emphasis being on the use of fresh, local ingredients to produce a tasty, healthy meal.

I have had the delightful experience of trying so many splendid foods from different locations during my global journeys. I have been fortunate to have had the opportunity to dazzle my taste buds with new culinary experiences in many situations. If I can find ways to dazzle my taste buds, so can you.

ADVENTURE SPORTS

What kind of adventures do you dream of having? Are you aware that your fantasies can become a reality?

Before reading any further, please know that I'm not a sports fanatic and that my style of travel doesn't usually involve sports. Let alone EXTREME sports.

I was petrified of skydiving. The thought of it frightened me beyond belief. Why would anyone want to jump out of an airplane in motion, suspended thousands of feet in the air, especially with someone strapped to their back? What if the parachute doesn't open? These are all questions I asked myself. Clearly, I thought, there was no need to do this. It made NO sense and would add NO value to my life. So, I made the logical choice, of course.

I decided to do it. Here's why.

The thought of skydiving showcased a fear of mine. Something I was scared to do. Yet at the same time, I saw an opportunity. A HUGE opportunity. One of EPIC proportions. I saw the opportunity to overcome this fear.

I signed up for the session with a Swiss guy I met, and off we went. We got suited up, ready to get into the plane, and were told that the conditions of the sky were not optimal and that we couldn't jump that day. There needs to be a level of clear visibility so that people can see through the clouds. Otherwise, the risk of jumping and colliding with an object exists. This uncertainty is a liability nightmare and an ethical no-no.

So the jump didn't happen that day. Phew! As it turned out, I wouldn't have to do it after all. I had been saved! I had the perfect excuse. Oh well, perhaps I'd do it at some other point in time...

...Except that's NOT how the story went. Why? Because I had learned from prior travels to build a level of flexibility into my trips for THIS EXACT REASON. Unplanned events occur and sometimes thwart your ability to embark on a major activity/part of your journey on the date and/or time planned. As it turned out, the Swiss guy and I still had the following day to do this crazy activity.

So back we went the next day. And we got suited up again. And we were once again told to get back into normal clothes because the weather was once again deemed unsuitable for a jump. It looked like I *actually wouldn't* do the jump after all...

But just as we were getting ready to go on with our day, the staff told us to get back into our gear and rounded us up to get into the small plane. And this time, we actually got on the plane. And the plane took off.

As we took off, I could feel my heart beating. I felt nervous. One minor detail I left out of the story is that we had elected to jump from the highest height offered – 19,000 feet.

It took what felt like forever to reach the elevation in which the instructor attached himself to the other couple in the plane's backs. Gave them instructions. Then out the open door they flew! My new friend and I looked at each other with that 'Uh oh, we're next' look. I was petrified with fear. What I would learn next would only amplify that emotion. We had only just surpassed 10,000 feet at that point in time. We still had another 9,000 feet to ascend before jumping. Feeling nauseous yet? I certainly was...

When the instructors gave us oxygen masks to wear, and told us to put them on, I knew things were getting pretty serious. As our altitude increased, breathing became a more onerous task.

And then the instructors strapped themselves to our backs and asked if we were ready. "I guess," replied my new friend. "Yeah!" I exclaimed. 'Well, sort of...uh, actually not really' was my afterthought that of course went unvocalized. I was feeling extremely nervous but had to come to grips with the fact that I would never be fully ready to jump out of a moving airplane at such a height. Who in their right mind would be? I knew I just needed to take action. To go for it. To face my fear head-on and conquer it.

And then my friend, with an instructor on his back, approached the open plane door, held onto it briefly, then out he flew! Gone. Into the wind. Literally. And I was next.

With the instructor attached to my back, I (we) approached the door. Out I looked. As scared as I've ever been in my life.

I couldn't turn back now. I held onto the sides of the plane. "3, 2, 1!" he yelled. And with a push, my grip slipped from the sides of the plane door, and I was in the air. Free. Falling. Freefalling. To my doom if the parachute decided it didn't feel like functioning that day. Oh well, that wasn't worth worrying about now. I had made my decision, and I was stuck with it.

Air whipped in my face so quickly that it was difficult to see. Hard to think. Hard to process what was happening. I tried to observe my surroundings. Amazing mountains. Ice. Green. But I was falling. Rapidly.

And all of a sudden, my body was yanked back upwards. The parachute had been pulled. It had worked! And I was no longer falling. I was floating. Gliding. Gently down to earth. Peacefully. Onto a field.

...and onto the field we landed. "How do you feel?" my instructor asked. "Great!" I replied. And that was the truth.

And I felt great, not just because endorphins were flowing throughout my body. I felt great for a deeper reason too. I felt great because I felt a burning sense of accomplishment. All of my life, I had been scared of heights. And that day, I committed to and took action on jumping out of a plane from 19,000 feet in the air with nothing but another human and parachutes attached to me.

I had just skydived over Franz Josef Glacier in New Zealand. It ranks as one of the top jumps in the world.

Let's make firm decisions and stick to them. Let's overcome our fears. Let's surmount life's challenges. And let's achieve extraordinary things.

We'll never be fully prepared to take on life's challenges, so we have to accept being only partially ready and commit to taking on the challenges anyway. The key is **TAKING ACTION**. Remember, nothing is impossible. Impossible things are just those that haven't been done yet.

Life is an adventure, and if I can experience it, so can you.

And if I can turn impossible into possible, so can you.

CHAPTER 5:

MANAGING YOUR FINANCES

You have received the motivation you need for your trip. You have received an overview of the benefits you will gain from your trip. But what about the money? That is what this chapter will focus on.

WHAT'S YOUR STYLE?

When thinking about taking a trip, it is important to consider whether income will be generated or not while you are away. If you have a stream of passive income (consistent revenue from a business, dividends, an annuity, pension payments, etc.) coming in while you are traveling, that will certainly be helpful. If you're planning on taking a long trip, you'll want to consider whether you will save first and then travel, work while traveling to earn income, do a save -- work -- save -- work pattern, or use another approach.

Chances are, if you're taking a two-week vacation from work, you won't plan on working during your trip. Your goal is likely to escape work for a bit and to relax.

However, if you're planning on traveling for a long time, you might consider working during your trip. I have met many a traveler who supported their trip financially in this manner.

The third category I like to describe as "rotational travelers." These are folks who work for a certain period of time, six months for example, save money, and then take off on a trip with the money they made. I have met Uber drivers, backpackers, and people from other walks of life who have employed this strategy successfully. They have done jobs such as picking berries, driving taxis, answering calls in a call center, and everything in between in order to fund their travels.

If going this route, it is particularly useful to look into the working holiday visa. There is usually an age limit (often 30), and the visa tends to be valid for a year or two and can sometimes be extended. Check which nations the country you are a citizen of has agreements with. The working holiday visa is a great way to experience a new country while earning the funds to support further travel!

I have personally made the decision to save enough money for my trips beforehand so that I could focus entirely on the experience while having it. ...Except for the time I became an epic DJ at a bar in Yangshuo, China for a few nights. ...Or that one night that I promoted for a club in Dali, China, received tons of free food and drinks, was invited on stage, and became an overnight celebrity. ...But I suppose the specifics of those stories belong in a separate book...

Which one of these trip styles would suit you best? Before reading any further, take out a sheet of paper and write down your answer to this question.

WORKING HARD, BEING RESPONSIBLE, AND SAVING MONEY

I often get asked, "but how do you afford to travel?" There is indeed a method behind the madness. And if I can utilize it effectively, perhaps you can too.

To be successful in financially planning for your trip, it's important to work hard, save money, and demonstrate responsibility.

My family and its values played a large role in the person I have become. I spent my high school years working hard at various local jobs and contributing financially to my family. I made payments to my mother for car insurance and other necessary things. I also gave her all the tips I had earned as a waiter at my sleepaway camp to offset the fee she had paid for me to attend. At graduation, not only was I given a custom-made book celebrating all of my hard work and achievements, but *I was also presented with the return of every cent I had ever paid my mother throughout those four years*, in original envelopes! I was taken aback, awestruck, overcome by a feeling that is hard to describe. A wave of emotions poured over me, all at the same time – gratitude, pride, happiness, amongst others. My mother had accomplished her goal of teaching me the valuable lesson that working hard and being responsible pays off.

I used the following strategies to save for my first extended, life-changing trip -- my 14.5-month excursion around Asia. I encourage you to study the below list and to see if any/all of these strategies might be helpful for you and applicable to your situation too:

- Studying hard academically, leading to the receipt of a scholarship for both undergraduate and graduate school
- Locating and living in a low-cost, high-value apartment complex during graduate school
- Working several jobs while studying
- Paying my credit cards off completely each month
- Using a basic phone
- Not spending on material/unnecessary things, rather saving for experiences

At the conclusion of my graduate degree, having received money back from my mom, having worked hard during my undergraduate and graduate years, having deliberately kept my expenses down, and having therefore saved a significant portion of money, I had developed the financial backbone necessary to support a trip. A trip that began immediately following my MBA. A MAJOR trip. An open-ended trip that ended up lasting for FOURTEEN-AND-A-HALF MONTHS OF MY LIFE.

At that point in time, I was fortunate to be in a situation where I was not strongly attached to anything or anyone back home. It was therefore relatively easy for me to take off on a journey in that regard. I acknowledge that not everyone is in a similar situation.

It is important to note as well that on this trip I was quite frugal wherever I could be. And for the most part, I spent on experiences. Not on material things. Being thoughtful about our purchases and deciding not to spend when we don't need to is a mindset, a skill, and an important part of the financial planning process.

Although I didn't realize it at the time, the knowledge gained during this and subsequent travel experiences would serve as the experience necessary to write this book, motivate others to live out their dreams, and allow travel to change their lives in so many phenomenal ways, just as it did mine!

PRE-TRIP AND BUDGETING

Now let's talk about how much you're actually going to need to save for your dream trip. This is going to vary person by person, trip by trip, and situation by situation. To help you to determine how much you'll need to save, I'm going to ask you a series of questions. Please write each one down on the same piece of paper you used to answer the question earlier in this chapter concerning your preferred style of trip. It will be to your benefit to answer the questions as truthfully as possible. Under each question, I'll provide suggestions, and under those, my own experiences:

Let's tackle some primary questions first:

- Why do you want to take this trip?

  o What is the motivation behind this trip? Is this a destination you've always dreamt of visiting and you want to do and experience as much as possible while there? Are you taking this trip to get away from life for a bit because you've just gotten out of a toxic relationship? Have you just quit or lost your job and decided to take a trip to find yourself and your meaning in this world?

    ▪ Most of my solo experiences, at least the longer ones, were taken as an opportunity for me to learn more about myself and about the world around me. I focused on taking in new perspectives, learning how to handle new situations, and challenging my beliefs and courses of action.

- Most trips I have taken with friends were aimed at relaxing together and building a stronger bond.

- Where do you want to go?

  - List all states, countries, continents, etc. Be as specific as possible. You can always cross items off of your list later. You don't need to have this list fully determined yet, especially if you're planning on taking a trip that is flexible and may not have a set end date yet. However, the more information you can provide now, the better.

    - I knew that I would visit Argentina, Brazil, Chile, Colombia and Peru on my South America Trip. The exact amount of time spent in each location was determined during the trip itself. Not before. My experiences in each location determined how long I wanted to, and ultimately ended up staying in each spot.

- What do you want to do there?

  - Consider any/all activities you want to do. Make a list of experiences you **definitely** want to have. Make another list of experiences you *may* want to have.

  - Determine which activities hold the most importance in your itinerary. Take the time to research the best way to go about doing them/the best and most reliable companies offering that experience. The last thing you'll want to do is have a subpar experience and feel the need to go back and re-do it.

o　If you are on an extended trip, use your time wisely. Plan during your down time (on bus/train rides, when it's too hot or cold outside to do anything else, etc.) so that your planning time doesn't have to cut into your activity time.

- I have had so many great experiences through travel -- from bar crawls in Europe to a coffee tour in Salento, Colombia.

- For my trip to Peru, I knew that Machu Picchu was a must-see. Seeing the Nazca Lines was a maybe. I ended up seeing Machu Picchu and not the Nazca Lines.

- During my 7-month world trip, while in one location I made plans to get to the next. I was efficient with my time. For example, I used peak sun time in Punta Cana, Dominican Republic – about 11:00 am to 3:00 pm local time – to plan out the next steps of my journey because the combination of fair skin and intense sun doesn't make for a comfortable or fun experience.

- When will you go?

o　Think about the time of year that you want to travel. Think about factors such as the conditions of the location you're planning on going to and how hotel/ accommodation, attraction, etc. costs may be affected (high season vs. low season).

- I chose to begin my world trip in the Caribbean so I could finish with that portion of it by early summer in order to avoid hurricane season.

Albeit providing great deals, this time of year also brings with it a great deal of risk. And I didn't want to take on that risk.

- I chose to visit the Salt Flats in Uyuni, Bolivia in March in an effort to maximize my chances of experiencing the magical mirror effect that allows one to appear as though they are walking in the sky. I was fortunate to experience it, along with the opportunity to take incredible perspective photos -- a win-win!

- Who will you go with?

  o Make sure you know this person well and that you have similar ideas and expectations as to what the journey will entail. It is helpful if everyone going has similar travel styles. Ideally, everyone going wants to do the same activities, is willing to compromise on some things, and/or is comfortable with the idea of splitting up during certain portions of the trip so that each person/group can do what he/she wants to do.

    - For almost all of my longer trips, I have traveled by myself. I have found that this provides me with independence and motivates me to meet more people.

    - I have also traveled with another person for an extended amount of time. We decided to do some things together, and other things separately. Sharing experiences together allowed for bonding to take place. Having our own space and being able to discuss the activities we experienced separately was refreshing.

- How long will you go for?

  o Be as specific as possible here. If you don't have a clear answer on this just yet, don't worry. Write down the estimated length of your trip.

    - My trips have ranged from a long weekend with friends to a 14.5-month excursion throughout Asia.

- What fixed costs do you have?

  o Time to take a deep dive into your finances and analyze all the things you're paying for, and see if there's a way not to (at least during your trip). Chances are you'll have a few of these. You'll still want to have a phone. One with the capacity to function globally and one that does not charge roaming fees is ideal!

  o The other big cost most people have is rent or their mortgage. Consider subletting your apartment out if allowed. If you own your home, you can consider a home swap agreement/timeshare situation. Or rent your place out while you're away.

    - Before my trip around Asia, I eliminated rent and storage costs by choosing not to renew my apartment lease and storing my belongings at a friend's house.

    - For several of my extended trips, I have taken the time to find an adequate replacement for my spot in my apartment.

Now let's now consider the following questions:

- What documents do you need for where you want to go?

  ○ You need to ensure you have all the necessary documents before traveling. This will usually involve having a passport for international travel or at least one form of government-issued identification if staying domestic.

  ○ Depending on your nationality and your desired destination, you may be subject to certain restrictions. And you may need to obtain a visa. **Give yourself MORE TIME than you think you'll need to obtain it to account for unanticipated events/delays occurring**.

  ○ Embassies and Consulates of the countries you are aiming to go to will likely be where you'll handle the visa process. Though you may be able to handle it entirely online.

  ○ Make sure you know how long you are able to legally stay in the countries you are planning to visit. Tourist visas often permit a stay of 1-3 months.

    ▪ My experience with obtaining a Chinese visa was painstakingly long and riddled with uncertainty. By giving myself more than enough time to work through the process, I was able to successfully secure this visa before my trip.

    ▪ During my Asia trip, I had to visit the Indian consulate in Hong Kong to obtain my Indian visa. I gave myself enough time to research the process and obtain the visa before needing to leave Hong Kong.

- Are there any health-related concerns you should be aware of?

  - Take the time to research the situation in the location you are planning on visiting. Are there any diseases that you need to be aware of? If so, can you protect against them in some way? Are vaccinations available and/or required for entry? If so, how effective are they and for how long will you stay protected? What level of risk would you be taking by going there, and are you comfortable with that level of risk? Does the specific location you are going to within the country have the same risk as other areas of the country? Can you avoid your exposure to the disease/s in any way? If not, are you able to mitigate your risk of contracting it somehow?

    - I received the Japanese Encephalitis Vaccine before heading to Japan in thinking that I might make it out into the Japanese countryside where the virus was prevalent.

    - When I traveled throughout Kenya, I took antimalarial medication to prevent malaria. Additionally, I took precautions to avoid mosquito bites, such as wearing pants and long-sleeved clothing, consistently applying mosquito repellent, and sleeping with a mosquito net hovering over my bed.

- How do you plan to get to your (initial or only) location?

  - Consider here that airline miles could work to your advantage and offer great value. Some credit cards offer great travel perks, so do your research and see which option (if any) will suit your situation best. Be aware of annual fees on credit cards.

- Most of my experiences have been international so I have usually traveled via airplane to arrive at the first destination of my trip.

- Where do you plan to stay on your travels?

  o Will you stay in hostels? Budget hotels? Luxury hotels? Private room or shared? What level of socialization/interaction do you want to have with others? Do you have a friend, family member, or someone else you could stay with? Do you plan on doing a room/apartment share? Will you be camping out somewhere? Is the neighborhood you are thinking to stay in safe? If not, is there another part of town you could stay in? (Your preferences may vary based on many factors such as your age, length of your trip, who's going, etc.)

    - In Europe, I stayed in a fair amount of hostels.

    - In Asia I stayed with a good amount of people.

    - On my world trip, I mainly stayed in budget hotels.

    - My style during my 20s tended to lean toward staying with people and in shared rooms in hostels.

    - Into my early 30s, I have been staying more in private rooms in hostels and in budget hotels.

- How and where do you plan on eating?

  - You don't need to plan out every meal (unless you have a very limited time for your trip and/or have certain popular places you'd like to eat at that may require a reservation way in advance). Think about whether you're mainly going to be eating fast/street food, eating at cafes, having fine dining experiences, going on food tours, etc.

  - There's no harm in combining some of the above options. You'll likely find yourself leaning toward one of the above more than others. Think about which one that may be.

    - I tend to do a mix of most of the above. I generally stick with cheaper options. I like to enjoy the local cuisine to better understand the culture of the place I am in and because I love trying new foods!

    - On our US Road Trip, my friend and I discovered a chain called Biscuit Head in Asheville, North Carolina, and boy was that a treat! We deemed it to be the best restaurant that we visited during our travels throughout 32 states! That's impressive!

    - In India, I was able to limit my food budget to, on average, somewhere between one and four US dollars per day due to the price of the items, but more so due to the bargaining skills I had honed throughout my time there.

That was a lot, I know! Take a deep breath. At this point, if you haven't already done so, please go back to the questions above, write them down on your piece of paper, and write down your answers.

Once you have the answers to the above questions, you can begin doing research into how much the flights, accommodations, activities, etc. will cost. Adding in factors such as where you'll go and when, how many people will be going, how long you think you'll travel for, etc. will allow you to put a budget together. List the name of each anticipated cost in its own row, using a separate column to house the cost of each corresponding item. You may find it helpful to group your costs into categories (flights, accommodations, activities, etc.). You may also decide you want to total up the entire trip cost, total up costs by day, by person, or by a combination of these elements. You can add formulas into your spreadsheet to help you accomplish this. Play around with whatever will work best for you and all parties involved in your trip in order to help everyone involved best understand the financial impact the trip will have on them.

Bonus Tip: Create a flex/emergency budget. From my experience, there is almost ALWAYS something that happens or doesn't go as planned, so it is wise to consider putting aside extra money for those "just-in-case" situations and the unknown. Just how much you put aside will also depend on the above factors (how long you're going for, how many people are going, etc.)

POST-TRIP

Great, so you've decided what kind of trip you're going to take, have answered the aforementioned questions, have hammered out a budget, and are good to go, right?! WRONG. The final thing to think about before your trip, ironically enough, is what your situation AFTER the trip is going to look like.

If you're taking a two-week vacation from your job. This question probably won't require much thought. You'll likely settle back into your workspace and will continue to receive a steady salary.

But what if that ISN'T your situation? What if you've recently quit your job and feel you're taking a HUGE risk? You're about to travel the world, yet you don't have any money coming in at the moment, and you don't know what you'll do when your trip is over. Well, THAT'S TOTALLY FINE! "What?! Really Rob?!" you may ask.

YES! I'VE DONE THIS MORE THAN A FEW TIMES IN MY LIFE AND I CAN CONFIDENTLY TELL YOU THAT WITH ENOUGH PRIOR AND PROPER PLANNING, YOU WILL BE FINE.

If you won't be returning to a salary (from a full-time or part-time job), ask yourself the following questions:

- Where do you plan on ending up after your trip?

  o This may be the country, state, city, town, province, etc. that you were born in and/or currently reside in, but it could also be somewhere entirely different. Maybe it's the final destination of your trip. Maybe the first. Maybe it's somewhere you don't plan on visiting during your trip.

  o Think about living arrangements. Do you have a home or apartment to return to? Will you stay with your parents? Your partner? On a friend's couch temporarily?

  o Be flexible. Be open-minded. You never know what opportunities life may bring your way, so keep your eyes open while traveling. And ALWAYS.

- After many of my trips, I returned to my apartment in The United States.

- Shortly after my Asia trip concluded, I received an opportunity to work in Japan. I took it and ended up working in Tokyo for three years!

- 1) What do you plan on doing after you arrive to your final destination? 2) How will you accomplish that? 3) How will your plan generate income if you are not returning to an established job?

  o Think this one through thoroughly. Do you have a source of recurring income coming in?

  o Do your best to put together a concrete action plan of *exactly* what you will do after your trip. Will you work for yourself or for someone else? In what industry? If working for yourself, what products/services will you sell? Is there a need for it/them in the marketplace? How will this idea generate money? How will you test this idea to ensure it does? If looking for work, will you begin looking before, during, or after your trip?

  o Be as specific as possible. Write down your goal/s. Write a detailed plan on how you will accomplish it/ them. Include a target date by which each goal will be achieved.

  o Break your goal/s up into smaller, actionable steps. This will make it/them seem less overwhelming, and will allow you to celebrate small wins along the way!

    - After my Europe Trip and going into the second year of my MBA, I knew that I would be returning to multiple income streams.

Bonus Tip: Set aside money to pay for your living expenses (food, rent/mortgage, etc.) for at least six months. This will help ease the transition between the end of your trip and what you decide to do after it. And you won't be financially stressed and have to dig into savings, go into debt, beg others for money, etc. while you work on putting your post-trip plans into action!

Now that you've got an idea of what to think about when planning your trip, let's get into the fun stuff! **IT'S TIME TO LEARN HOW TRAVEL WILL CHANGE YOUR LIFE!**

CHAPTER 6:

PUTTING PLANS INTO PLACE

Travel will inevitably make you a more organized person. Learning how to make solid decisions, effectively plan your trip, and efficiently manage your time are all byproducts you will realize.

Let's look at the steps involved in this process:

1) We want to make good decisions and plans that will allow us to maximize our time doing the activities we know we want to do

2) We want to allow for some degree of flexibility for people we meet and/or places or activities we may discover along the way

3) We want to ensure we understand/learn how to best utilize our time

DECISION-MAKING

Are you a good decision maker? That is an important question to ask yourself. It is important for us to decide how we want to spend our time since it is limited on our trips. And on this Earth. Deciding on the activities and experiences you **absolutely** want to have is an important element of trip planning.

If you are more of a planner, you may want to arrange everything in a very organized and detailed fashion. You might want to plan well in advance in order to secure better financial deals and eliminate the need to think about the trip any further, knowing that everything is ready to go! You can contact a travel agent to handle some or all of your booking arrangements if you don't want to deal with the actual travel logistics yourself.

If you are more spontaneous, you may prefer to wait until closer to the time of your trip to book anything. You might also decide to book the trip one leg at a time as you may not know exactly how long you will want to stay in one place for. Doing this will account for people you may meet or information you may gain along the way that could influence your desired plans.

Once you've decided on the experiences you KNOW you want to have, stick to them and work them into your plan. You'll learn to work around these definite activities that are *not* flexible, sprinkling your "maybe" activities in around them. For example, during our United States Road Trip, Lincoln and I signed up for a tour of Antelope Canyon in Arizona. This was an activity we knew we wanted to do, and clearly a popular one. In accordance with our timing, there was only one available time slot, on one day, that could have worked for us. So we booked the only time slot available and planned our other activities in Arizona around the Antelope Canyon experience.

FLEXIBILITY

Leaving room in my travels for flexibility has been an incredibly rewarding experience. As mentioned in a prior chapter, it has allowed me to have a second chance at the skydive I wouldn't have been able to have done had I only left myself one day to do it. It has allowed me to meet local people and try new things that I hadn't imagined doing or trying before arriving at the location.

For example, I hadn't planned on going anywhere near Tibet while in China, however I met an Israeli girl who raved about the value of going further west than just Shangri-La (where I was at that time). That by doing so, I would **really** get a sense for the "Wild West" of China, and that I would start to see and feel what Tibet was like. I hadn't planned on going any further west than Shangri-La, but through our brief conversation, I became convinced.

I was fortunate enough to find a guy in my hostel who said he would be hitchhiking to get a ride further west...toward Tibet. I had heard that the Chinese police were detaining Chinese people and tourists alike within, going to, and coming from that region. I was hesitant...and I would go on to find out something very, very interesting during this journey...

The guy from my hostel asked if I wanted to join him in hitchhiking for a ride further west. 'How absurd, I could be kidnapped!' I thought. So, naturally, I agreed to partake in this seemingly ridiculous activity with him. We tried for a while without success.

Until we came across a group of Tibetan guys who scooped us up. And off we went! I was nervous at first, but my new friend assured me that this type of situation was not only safe, but actually a very common occurrence in this region.

The crystal-clear waters of the small, blue, untouched ponds I saw throughout the mountains visible from the rarely traveled roads we drove on in that part of China will stay ingrained in my mind forever. No garbage. No pollution. No other people. Just us. And nature. Incredible. Rare. Breathtaking. Surreal.

This group of friends ended up bringing my friend from the hostel and me all the way to a town named Litang, with an overnight stop in a smaller city called De Rong along the way. The Tibetan guys paid for all of my food, the karaoke experience they insisted we have, and even my hotel room! Absurd! These guys even ended up bringing me back from Litang to Shangri-La several days later. All the while, REFUSING TO ACCEPT ANY MONEY FROM ME WHATSOEVER.

Litang felt to me like the "Wild West" of China. What Americans would see as "cowboy" attire was rampant. The norm actually. From boots to hats – the whole outfit. Everyone who saw me would smile and wave and would want to engage in conversation. They were curious about me. And about what I was doing in their town. I too was curious about them. And to know more about their town. Its people. Its history. Its culture. Its secrets.

The experiences I had with this group of people were epic. We had an amazing time together. With our arms wrapped around one another within mere hours of becoming acquainted, drinking (appropriate beverages of course…), and singing Tibetan songs (the likes of which I knew not a THING about), clearly rapport had been established. Quickly. But how was that even possible?! WE DIDN'T SPEAK A COMMON LANGUAGE! We found

a flight however, being late will *never* be a good idea. In fact, you may miss check-in. And the opportunity to drop your luggage off. Or have to wait in an astronomically long security and/or customs line -- all of which will add additional stress to your journey and could even cause you to miss your flight.

During my trip around the world, I found that even though I thought I had left early enough to catch my flight out of London, I still arrived at Gatwick Airport to find some things I hadn't realized beforehand. The line for the security check was enormous, and I would later learn just how large the airport was and how far my gate was from the security checkpoint. This is also something you'll want to check -- where is your gate and how far from the security check is it? Do you need to get to another part of the airport? If so, can you walk there? If not, is there some form of transportation available? And if so, how frequently does it run?

Fortunately, I was able to pay a bit extra at Gatwick Airport to gain access to the expedited security check line. Had I not found out about that, and had I not opted to go for it, I would not have made it to my gate on time and surely would have missed my flight. Although physically exhausted from running, I had arrived just before the cabin doors closed. And fortunately, the staff was still letting people onto the plane at that point.

That incident taught me how to better manage my time -- I would always recommend arriving at the airport **before you think you need to be there** to account for unknown factors. **I would advise you to arrive at the airport a minimum of three hours in advance for international flights and a minimum of two hours before domestic flights.**

Perhaps you forgot your passport at home and must return to retrieve it. Perhaps the line for passport control is extremely long. Perhaps the airport is a lot larger in person than it appears to be on the map. Perhaps the security check line is really long. Perhaps you get randomly selected at the security checkpoint to be the lucky receiver of a more in-depth search. All of these situations have happened to me at least once in my life. And all of these situations could happen to you. *Give yourself more time than you think you'll need at the airport! Always!*

Pack as light as you can. If possible, avoid bringing and checking a large bag or suitcase. I know this sounds crazy. Maybe it is. But perhaps the lighter we pack, the lighter our spirits become...

That's what I did for my 7-month trip around the world, and I never had to account for the time associated with checking a bag. Check in online (with a web browser or app) beforehand if you can to save time. Because I didn't check any bags, I never had to worry about the airline losing my baggage (and all of the headaches associated with that situation) or wasting time waiting at the baggage carousel for my luggage. **Maximize and be efficient with your time.**

Concerning time, we must realize that **no matter how well we plan our trips, and even our lives, we will never have enough time to do EVERYTHING.** This is an extremely important concept to realize and to internalize. And it took a lot of time for it to soak in for me. Being cognizant that our time is finite should enable us to better focus on how we want to allocate it. Since we will never have time to do everything, we must pick and choose what is important to us. Some travelers may want to cram as many sights and activities into their trip as possible. I used to travel that way. And there isn't anything wrong with traveling that way. Especially if one's trip time is quite limited. Through my travel experiences, I have come to realize the value of focusing on less but more in-depth and intentional experiences that align with the goal/s of my trip.

How would you like to travel? Would you prefer to fit as much as possible into your itinerary? Would you prefer to plan only a few but very meaningful activities? Somewhere in between perhaps? Will you leave yourself with flexible time to wander around and explore the place you're in?

In thinking about how best to spend your time, a travel coach can help you find and set the intention and meaning behind your trip/s. He or she can assist you in setting a goal and sticking to it. A mantra, or personal statement, may be created and used to guide you on your journey and help keep you focused on attaining your goal. A travel coach can also keep you on track when you return from your trip, holding you accountable for incorporating how you changed and what you learned on your travels into your daily life.

CHAPTER 7:

LEARNING MORE ABOUT YOURSELF

As mentioned previously, I can confidently say that **I have gained more practical knowledge from my outside-of-the-classroom (real world) learning experiences than I have from all classrooms and work environments I have been in combined.** That's a powerful statement. Take a moment to read it again.

I'll discuss the importance of learning about the world, but in this chapter, I'm going to focus on what you will learn about *yourself* through travel. The answer? A LOT.

Travel has allowed me to connect with my heritage. I have been fortunate enough to have had the opportunity to visit Israel twice! One trip provided me with more of a general overview of and introduction to the country, its history, and to Judaism. The other trip was more focused on the Jewish Religion and learning. On both occasions, I had the opportunity to see many places, partake in many interesting activities, and meet many interesting people. Perhaps most interestingly, I had the opportunity to connect with my roots.

And speaking of roots, my family immigrated to The United States from various parts of Eastern Europe generations ago. Vienna, Austria being one of those places. I therefore knew that on my Europe trip I would need to put that city on my list. And I did. And I went. And I discovered magic. Magic? YES. Walking on the very same streets that my ancestors had traversed oh so many years ago generated a magical feeling inside of me. Part of me had come from this city. I was a part of it. I am *still* a part of it. My soul will **ALWAYS BE** a part of it, just as it will always be a part of me.

Throughout my Europe travels, I also discovered Maisel's Weisse, a German beer brand including my last name -- how cool! Could our family be related to this company? Growing up, I had seen mugs, glasses, and bottles with this same logo in our house. Maybe, just maybe, I had stumbled across something...

Are you curious about your heritage? Would that curiosity influence where you would want to travel to?

In addition, travel has been a vehicle for me to be able to better understand and challenge my thoughts and actions. Perhaps you may find the same.

OUR THOUGHTS

Why do you think the way you do?

I believe that the answer to this question is a combination of factors: your personality, your genetics, the society you were raised in, the experiences you have had, and the list goes on.

Travel can help you think in new ways about situations and experiences you've had.

Growing up in The United States, I was raised to believe that stating my opinion and being expressive about my feelings is a good quality. I believed this and didn't question it. But through my experiences in another culture, I would come to question this.

After living in Japan for three years, I learned that being honest with my words if my thinking wasn't in agreement with that of the group, could have some negative consequences. It seemed to me that in Japanese culture, it is more acceptable to agree with the group, even if you're asked your opinion and you disagree. I realized that answering a question honestly, if it involves negativity or a potentially harsh or painful truth, can do more harm than good in Japanese culture. It can cause the other person/people to lose face, which is a grave mistake to make in relationship-building. Especially in Japan. I also found that delivering a bold and honest opinion, particularly one that varies drastically from that of the group, can and almost does make the person offering the opinion appear as an outlier rather than a helpful part of the group. Being bold and opinionated got me nowhere in that society. But learning to be a cohesive and synergistic catalyst in a group did.

I therefore began to question the value of this idea -- of being expressive of my emotions and opinions. I came to the conclusion that it can be beneficial or detrimental, depending on the context of the interaction, and pertaining largely to the society in which one lives and/or is interacting. Living in Japan and being away from The United States helped me understand how norms can vary greatly between cultures. Perhaps I had initially valued being expressive so much, largely because of how much United States society (overall) values that trait. And depending on the experiences I will have moving forward, perhaps my view on this issue will shift yet again. I believe it is important that we remain open-minded, receptive to new ideas and concepts, and able to be persuaded by experiential learning and the presentation of compelling arguments.

By seeing other ways of thinking, we are naturally forced to think about our own ways of seeing the world and doing things. Is there an experience you had that challenged you to adapt the way you think? Did you gain a new perspective?

During my time in India, my perception of the value of honesty would be challenged even further. It would be stretched in new ways and to new limits...

While there, I had numerous encounters with people who seemed to be after my money or trying to scam me in some way, shape, or form. I often felt like a walking ATM, and it appeared that just about everyone who came up to me wanted something from me. This was my experience. And that is how I felt at the time. Yet I am aware that we each have our own experiences. Our own ways of processing them. And our own feelings that result from them.

In one instance, a guy told me he liked my shoes. I ignored him, yet he kept asking why I was ignoring him so I finally turned around and told him about my experiences in India and my feelings based upon the interactions I had had. He insisted that he was not interested in my money and that he was nothing like the other people I had encountered.

A friendship began to ensue between us. He took me for chai, to see where his relatives work, and even offered to take me out for a night on the town. In what I was told was an unfortunate accident that had occurred with someone in his family, we would have to shift plans from going to a nightclub to having a coffee. And his "uncle" would come. The "uncle" ended up buying me coffee and proceeding to discuss his business of exporting gems. He mentioned it was quite the lucrative business. I became a bit uncomfortable when personal questions such as what kind of credit cards I use began to be asked. When offered to join this venture, I told the pair that I would think about the proposition and get back to them.

After arriving home later that night, I discussed what had happened with my travel partner. He quickly leafed through his version of the guide book he had for Southeast Asia and identified what they had been attempting to get me to agree to as the infamous "gem scam." It is a common scheme where people (mainly tourists within India) are sent overseas with expensive gems and told not to declare the goods at the airport. If the gems are able to be sold, the person moving them can make very good money, however just transporting them over the border would result in a hefty profit by the transporter of said goods. The stakes of getting caught however are high, and the punishments far outweigh the rewards. Through this experience, I learned to be more cautious of who I trust.

That same night, my travel partner said he had met a local and was invited to that person's home. All of that guy's friends sat around in a circle while this guy played songs on the guitar, throwing in money all the while, and of course my friend partook in the activity. And came home with empty pockets. I then had the epiphany that after my friend had left, it was highly likely that all of this guy's other friends (if they really were friends, and not business associates) would take their money back. But my friend would never see his again. They would likely divide the money left by my travel partner up and call it a night. My perception was that this was a scam. Perhaps the perception of the locals was that this was just a simple game. A game that they had clearly won. But we wouldn't continue to lose the game forever. We would ultimately go on to learn how to beat them at their own game.

Back to Varanasi, India for a moment. The same place where my travel partner and I had been barraged by rocks and chased by a horde of people after the owl had been set free. When we first arrived in Varanasi, we gave our tuk-tuk driver the name and address of our guest house. We were headed to a very famous spot that was written about in guide books, and we therefore felt we could trust it. When we arrived, something seemed off. Upon double-checking the address, we realized we had been taken to a different place. A place that had a similar name. But a different address. A knockoff.

When we told our driver that we had been taken to the wrong place, he bobbled his head back and forth. When we asked him to take us to our actual destination, he bobbled his head back and forth once more and encouraged us to come inside the place we were already at, seemingly with the aim of having us stay there. We repeated that we wanted to go to the place we originally asked him to take us to. He bobbled his head again. I will discuss my experience with this interesting phenomenon – "the head bobble" – in more detail a bit further along in the book.

Perhaps our driver was having trouble understanding us. Perhaps he was getting a commission for bringing guests to THIS guest house and not the other one. He didn't end up taking us to the other guest house. He didn't end up getting paid either. We walked the rest of the way, asking people we encountered to guide us to the actual guest house. And after several confusing conversations, we ultimately arrived at the correct location. It was in that guest house that I would meet another lifelong friend. A gentleman I would go on to visit with both in his home country of Taiwan and in Japan.

After traveling around to more places and meeting other Indian people outside of India I began to see a new perspective. Yes, many people in the tourist industry I had interacted with attempted to charge me many times the Indian price for the good or service at hand. But they also had to survive. As a third-world economy, there is a lot of poverty in India. I began to better understand their thought processes and actions -- as survival tactics to fulfill their basic needs rather than as personal attacks on me. I also began to comprehend the intelligence behind some of the planned efforts and strategies employed. While I would not agree with many of the tactics locals used on me to separate me from my money, I am now better able to see both sides of the coin.

Has anything caused you to rethink how you view an event or situation that occurred in your life? If so, what was your initial perspective? How did your thinking shift? What is your new perspective on the matter? If you haven't experienced a shift in thinking or the ability to view an experience in a new light from travel yet, could you see how this could happen and how it could be valuable?

What you learn through travel has the ability to impact how you view **future** events too.

The story I mentioned earlier about my day with the elephants was very powerful. Learning about the cruelty included in many activities involving animals has caused me to think about some of my past experiences. Although at the time I was unaware of the impacts of some of these activities that I partook in, I am not happy that animals may have suffered because of them. Moreover, the experiences have caused me to re-evaluate which activities I want to partake in moving forward. With this elevated level of knowledge and awareness about animal cruelty and suffering, I believe I have a higher level of responsibility. My thought process in future decision-making pertaining to activities involving animals has been altered. Moving forward, I will be more careful to select and participate in activities that I know or believe there to be no animal cruelty involved. Because I don't want my money supporting that cause.

Has an experience you have had shaped your perspective on how you will approach future situations?

OUR ACTIONS

Travel also enables us to analyze our actions and **WHY** we do them.

Why do you behave the way you do?

I have found that we generally *aim* to act in ways that align with our beliefs and values. We *intend* to act in ways that allow us to accomplish our goals and desires. Actually, most of the time we just act out of habit and don't take the time to think about *why* we are actually doing what we are doing -- we are on autopilot. Our subconscious mind is in control.

At the Great Pyramids of Giza, I was constantly hassled with offers to ride camels and had learned the day before from someone who had been there that in these situations, the people offering the experience often quote one price before the ride, only to demand a higher one afterward.

Some folks within the pyramid complex asked to see my entrance ticket. Luckily, before entering the complex I had been advised by my guide to ignore these people because although they are posing as property workers, in fact, they are *not*. They ask to see tickets in order to gain access to them, only to demand a fee to return them to their rightful owner. I was able to avoid this whole ordeal by politely smiling and saying "no thank you," or by putting my hand up, waving it from side to side to indicate that I wasn't interested. I believe that in the past, I would have gotten annoyed at these occurrences. Many folks do, and some even let it ruin their entire experience. From my time at the Great Pyramids of Giza, I learned to ignore small annoyances in life and not let them detract from the bigger, better picture that we should be focusing our time, attention, and energy on.

One example of a way that travel changed my actions is as follows -- I noticed I used to ask many questions. I did this in an effort to show that I was interested in learning more about the person I was speaking with. By living in Japan, I learned that for the most part, a barrage of questions is not appreciated in that society. Asking too many questions tends to make Japanese people feel pressured. My experience is that folks in that society generally don't respond well to pressure and may therefore not even respond at all. An elevated level of caution should be exercised when asking questions to a Japanese person because questions can actually be seen as indirect complaints. With all of that in mind, I adapted my communication style radically during my three years living there. Living in a new society allowed me to challenge my traditional ways of acting.

I also realized that asking questions too personal in nature was not a good idea, so I learned to be culturally sensitive and creative. While working as an English teacher, I noticed that when asked what company they worked for, most of my students skirted the question or answered with a level of unease. As Japanese people tend to be pretty private, even that question can be seen as too invasive. I learned to ask *what kind of work* they did instead of *their company's name*. This way, the student could share whatever level of detail he/she was comfortable with. I found that with this approach, I got more favorable responses and was able to build a stronger rapport with my students.

In The United States, we generally pay a set price for an item and that's it. Item sold. Transaction done. The idea of walking into a store and buying something in this manner makes sense to most Americans. While I was in India, I noticed that if I did this, I would pay a much higher price than what the item was actually worth. I learned to haggle my way through transactions and would often begin to walk away from sellers with conviction until I heard things like "hello sir," and "okay, okay," in an effort to regain my attention and offer a lower price. I knew I needed to adapt my strategy and actions in order to be effective in that new environment. So I did. And if I can do it, so can you.

Have you had an experience that caused you to alter your actions?

CHAPTER 8:

CAUTION: IMMENSE GROWTH AHEAD

Travel will help you to grow both personally and professionally. I can say this because I have seen it happen to me. Let's break the growth down into the following two areas: personal growth and professional growth.

PERSONAL GROWTH

As the last chapter highlighted, one of the benefits of travel is its ability to shift your mindset and your actions.

There is a profound change that I made in my life, which is worthy of mention here. And I have travel to thank for providing me with the experiences and the courage to make that powerful change.

Growing up, I had eaten meat because it was what most people around me did. It was what my parents did. It was what my friends did. And therefore, at that age, it seemed to be okay, and ended up being what I did. As a child, after learning where meat, poultry, seafood, etc. come from, I felt uncomfortable and guilty eating these things. But I tolerated it due to social pressure and convenience. I hadn't encountered a place where these items weren't eaten. Until I arrived in India. Strongly influenced by my experiences there, I decided to set out on the vegetarian path I had always known was right for me.

The town of Amritsar in Northern India was the first place I visited in my life where not only is it abnormal to eat meat, you can't find it anywhere even if you wanted to. I was at ease concerning my diet for the first time in my life. I didn't have to be concerned about where I went to eat. I didn't have to ask anyone if vegetarian options were available. It was an amazing and uplifting feeling. I realized that albeit having different cultural norms and values, there are communities of people in this world who feel the same way as I do about this issue. And therefore, have similar eating habits. And I had found one of them. I had found my community. Far, far away from where I had grown up. In a place I never would have imagined.

Further, upon visiting Mumbai, India later that trip, I went to the meat market to better understand the process of how meat arrives on people's plates. There, I saw all types of carcasses baking in the sun, with an immeasurable amount of flies buzzing around them. The sight was grotesque. The stench, nauseating. My experiences in India reaffirmed how I had always felt about eating meat. And reinforced why I didn't want to.

Difficult and challenging experiences will cause you to become more emotionally mature. You will learn how to think through and handle difficult situations.

I saw a monkey with a chain around its neck (being used as a leash) in Varanasi, India. The monkey was entertaining people on the street and making the man money in the process. At first glance, I felt that what the man was doing was unethical and that he should find another way to earn a living. I felt sad witnessing and reflecting upon the situation.

This situation made me want to focus on what I can control in my life, and this is a helpful skill to have when it comes to decision-making. I saw a situation which seemed that there could be elements of cruelty and potential abuse involved. And this experience made me want to make changes in my life to stop the abuse of animals. Regardless of whether or not cruelty was involved in this scenario. I might not be able to stop animal cruelty worldwide by myself, but I can certainly share my thoughts with others. And I can choose where I want to spend my money, and therefore decide not to support activities or groups that I believe are involved in said cruelty. I also have the power to decide what food and products I buy and consume. And to inform myself of the processes that occur in order for them to arrive on the shelves of stores.

Do you generally focus on what you can control in life or on what you can't?

Upon further analysis, I tried to understand why this man with the monkey was doing what he was doing. I tried to put myself in his shoes. Was he an evil person, or was this perhaps the only way he knew how to make a living? Perhaps he saw nothing unethical about what he was doing and/or tried other forms of work in the past that did not work out. Maybe the monkey was only chained by the neck when they were entertaining, and they actually lived somewhere else where the monkey was free to roam. Those were facts I did not have, and therefore things I thought best not to assume.

Through this experience and others, travel has taught me to be a more emotionally mature person. I have learned how to better regulate my emotions. This is often easier said than done. Especially when you are confronted with a situation, often within the context of a relatively unknown culture, that, at first glance, seems to challenge your values. And appears unethical. Or even outright cruel.

I have learned to analyze situations from various angles, even when they initially appear to be very disturbing. And to try to remain as objective as possible. By keeping an open mind, I am able to explore multiple perspectives of an issue or situation, analyze it, and make a decision if need be, based upon the facts and the evidence available to me.

Are you able to regulate your emotions?

Travel will teach you to stand up for yourself and what you believe in and to do the right thing.

On a bus journey between towns in India, I had an interaction with a gentleman who was collecting payment for the bus tickets. A sly guy. Lanky. Owner of a mustache. He had a smooth, used car salesman demeanor to him.

He told me I would need to purchase two tickets. One for me and one for my backpack. But my backpack wasn't taking up a seat.

And it wasn't preventing anyone from sitting down. And I told him that. And he told me I still needed to pay.

Growing up, I had been fearful of speaking up for myself. Often saying nothing when I felt I should have said something. Not comfortable with and scared of confrontation, I would choose to avoid problems, and as a result, they would fester. Instead of expressing my beliefs and feelings, I would ignore the issues at hand, leading to feelings of anger and resentment later on. I didn't like handling things in this manner. And I knew I wanted to become a more active and engaged problem solver. I knew I wanted to confront situations when necessary, finding the balance of when to walk away and when (and how) to "go to battle." Going through many of the experiences I went through on this trip, particularly in India, gave me the courage to make that change. This experience in particular stands out in my mind.

I sensed I was being treated unfairly but didn't want to jump to any conclusions. So I subtly and quietly struck up a conversation with the woman sitting next to me. I informed her of my dilemma. I asked her whether or not people on the bus ever have to pay for a second ticket due to their luggage. She reluctantly told me they didn't. She didn't want to get caught in the crossfire -- but by being honest with me, she gave me all the ammunition I needed to fire back. And to fire back with force. In this case, a battle would be necessary, and I was prepared to fight it.

I informed the gentleman collecting money that I was aware that nobody else had to pay for an extra ticket for their luggage. And he persisted that I still had to pay for mine. This was when I became more assertive. And he began to listen more intently. He bobbled his head from left to right several times, and after a few minutes of back and forth, and with reluctance, he agreed to accept payment for just one seat.

So I agreed to pay for one seat. And I did. But I didn't have small bills. And he didn't give me any change.

And so I brought Assertive Rob back out again in order to receive my change. I persisted. And he relented. Grudgingly forking over what was owed to me.

I was not fighting to save money. I was fighting to defend my principles. I was fighting to defend my values, my morals, and my beliefs. I was fighting to defend what was right. This was about sticking up for myself. Addressing a problem instead of letting someone trample over me. And most importantly, STANDING UP FOR WHAT IS RIGHT. If this happened to me, I was convinced that it could and likely would happen to others as well. And someone needed to put their foot down. So that day. On that bus. I did. **Changing the world starts with us -- we ALL have the power to do it. One interaction at a time.**

Travel will also teach you to become more flexible and resourceful.

I was confronted with a situation in Stockholm, Sweden, where I had a place to sleep lined up, and my host canceled on me last-minute. With the new information of not having a place to sleep for the night, I needed to search for a place to lay my head, with little time to find one.

It was summertime, and most of the hostels were at full capacity. Instead of panicking, I took off on the town with my backpack to explore! I wandered around the city, searching, chatting, and inquiring. And by doing so, I was able to better understand the city and its people and ultimately find a bed to sleep on for the night! Exploring a city with an open mind, speaking to people from that region or to those who have moved there from somewhere else, and getting recommendations from them is a superb way to dive deeper into the culture and history of the place you are in, allowing for a truly local experience!

Could you see how travel experiences can make you a more flexible and resourceful person?

Travel will show you aspects of your character and personality that may have been unknown to you prior. Or it may magnify certain elements that you were already aware of.

Travel will teach you about your strengths and to appreciate yourself for who you are.

I learned that my outgoing personality, my upbeat and positive attitude, and my smile are all major strengths. I learned to love my willingness and eagerness to engage with people from other cultures. And to travel to places freely and spontaneously. To strike up interesting conversations at a moment's notice. And to share stories, knowledge, experiences, etc. with people. **I realized how powerful my curiosity to learn about others is, so much so that it motivated me to travel to many locations, learn about new cultures, and learn new languages to be able to better connect with a plethora of diverse communities and people worldwide!**

What are some of your strengths? What are some things you love about yourself and why?

Through travel, I also identified weaknesses. It is important to figure out what those traits, habits, and behaviors are so that we can be aware of them and work on changing them if we decide we want to. I realized I was, at times, too focused on one minor element of a trip or an experience. I therefore started focusing more on the bigger picture and less on the minutia.

What are your weaknesses? Is there a way you could improve on them if you wanted to?

Traveling, especially if you go on your own, will teach you to be a more independent person, which in turn will boost your confidence.

This is because when you're traveling on your own, you will often have nobody to rely on but yourself. You'll be the one making all the decisions -- which locations to visit, what foods to try, which activities to partake in, and so on. This is a powerful and important skill to develop. And travel will help you develop it!

Travel teaches you to be grateful.

During my semester abroad in Buenos Aires, Argentina I took a trip to Colonia, Uruguay. Never before had I experienced that kind of poverty. Horses trotting along through the streets carrying troves of garbage – "garbage caballos (horses)" as I named them in Spanglish. People on the streets in ragged and tattered clothing. Dimly lit areas all over the city due to spotty electricity. From experiencing what these people *didn't have*, I felt so fortunate for what I *did have*. And I have carried this deep appreciation for all of the amazing people and things in my life into the present, and into the future it shall come with me as well.

When I was traveling through Cambodia, I stumbled across a local organization that was not only bringing clean and drinkable water to villages but was also providing the children with the opportunity to learn about various subjects. I took part in the learning by teaching English for the day. Seeing how much these children appreciated the time and effort of the volunteers was incredibly rewarding internally and produced in me a feeling of deep satisfaction. This experience pushed me to be a more kind, considerate, and caring person. It also reinforced my gratitude for all of the many amazing things and people I have in my life. If you are reading this (or listening to this), I am grateful for YOU!

Through my travels, I have encountered many people and communities living in poverty. Although one might surmise that those in such a situation would be miserable, I have, in almost every situation, found the *complete opposite* to be true. How is it possible that people who have so little can be so happy? Perhaps material objects can cause short-term happiness, but not the long-lasting kind. Perhaps without the distractions of so many material things, we will have the opportunity to appreciate and focus on ourselves and our relationships with others. Perhaps material objects can enhance happiness but don't produce it. Perhaps happiness comes from within and from our connections with others and the world, rather than emanating from the screens of our technological gadgets.

Travel also teaches you to be resilient.

During my semester abroad, through the front door they came. Making a ruckus, but nothing out of the ordinary for a weekend. I was used to my host brother having friends over, and although they were a bit noisy, the situation was tolerable. I was a bit annoyed since I was on a video call, but there were more important things to be concerned about in life...like life itself as I would come to learn...

When I took a peek into the other room, I didn't see a large group of my host brother's friends. I didn't even see the host brother I was expecting to see. Instead, I spotted my youngest host brother. And my host sister. And they weren't with a large group of friends. Only two. Who appeared to be a fair amount older than them. And who were, for some reason, armed. With guns. And who, I realized, *weren't friends of any of my host siblings*. They were strangers. And they were in the apartment for a reason. One drastically different than friendship.

A group of armed robbers had been waiting in their car in the early hours of the morning when the perfect victim arrived -- my host sister. She had just gotten dropped off from a birthday party. But at the corner of the street rather than in front of the main entrance of the building. In the minuscule amount of time that it took her to exit the car and arrive at the front door of the building, these men took swift action. My host sister found herself with a weapon pointed at her head, being instructed by two unfamiliar men to take her into our building and up to our apartment.

Once I realized what was going on, I became stricken with sheer terror. I was confronted by a man who tapped me under the eye with the handle of his pistol asking, "Me entendés?" (Do you understand me?) I was unable to respond. My vocal cords were frozen. Immobilized by fear, I had lost the ability to speak. I felt weak. Powerless. Thoughts of heroic actions I could have taken only came to me in hindsight. In that moment, my body allowed me to do absolutely nothing. Except to go under the bed as I was instructed. And luckily, while under there, I failed at my attempt to contact the police. For if I had had the right number on hand, and should the call have gone through, the intruders likely would have heard my conversation. *That phone call could have cost me my life.*

Fortunately, the youngest of my host brothers handled the situation quite well. I was amazed at the courage of this young boy and how calmly he interacted with the armed robbers. He gave them the material objects they wanted, and eventually, off they went. Perhaps never thinking of the effect they had on me or on my host family. Likely never thinking of the effect they had on me or on my host family. Almost certainly never thinking of the effect they had on me or on my host family.

They took my phone, my camera, and my laptop. They also instilled fear deep inside my heart. After that incident, I would walk down the street, turning around to check if anyone was behind or near me at all times. I was scared during the day. I felt scared when evening rolled around. I worried it would happen again. And I didn't have my main support system -- my family and close friends -- around to listen or to talk to in person. This was one of the most challenging times in my life.

But I was grateful. Grateful to be alive. And having had my life flash before my eyes, I had learned how quickly life can vanish. And from that point forward, I began to appreciate life more. What a phenomenal learning experience this was! Rather than viewing this memory as a traumatic event, over time I would come to see this incident as a blessing in disguise.

Although not an easy or quick process, fortunately, my fear would ultimately dissipate. I was thankful that most of the technology had been removed from my life during that incident so that I could spend more of my time engaged in and interacting with Argentine society and with my host family. I ended up having a more enriching experience, interacting with more of the local people and culture, and creating stronger relationships because of it.

I could have let that terrifying experience ruin my spirit to travel. But I didn't. Instead, I used it as motivation to explore the world at a highly accelerated pace. That experience taught me that life's clock will indeed run out at some point in time, and since the majority of us don't know when that point in time will be, we may as well do and see everything we want to while we are able to.

Now that you've heard my story, I'd like you to think about your own. Have you encountered any large challenges in life? How did you approach the situation/s? What did you learn from it/them?

PROFESSIONAL GROWTH

While traveling, you will grow as a professional as well, gaining skills that are relevant to the workplace.

You'll learn new ways to handle situations and will gain the ability to look at problems from new perspectives.

My experience living and working in Japan provided me with insights on how to look at situations in a new light. Before this experience, my American upbringing was the sole force dictating how I viewed my work. I held the perspective that independent thoughts and actions are good. Coming to definite decisions is necessary. Having solid stances on arguments and providing reasoning for why we support them is important. Providing and receiving critical feedback is crucial.

But that's just one approach. And it doesn't always work.

While living in Japan, I realized that in order to be efficient and effective, I would need to reprogram myself to use a different operating system. In Japan, my ability to show compassion and thoughtfulness for others was valuable. My ability to interact with a group cohesively in order to create synergy was powerful and important. NOT my ability to be an individual. NOT my ability to be strongly opinionated. NOT my ability to act independently. Saving face or preserving one's reputation, I learned, is paramount.

Experiencing and understanding these differing work styles has allowed me to better understand the world around me by being able to assess situations from new angles. I have learned how to adapt to specific situations and how to tailor my actions accordingly, depending on where I am in the world and/or what group of people I am communicating and working with. The term "cultural chameleon" may be relevant here. Would you like to learn how to be a cultural chameleon too?

In the Japanese middle school in which I worked, I co-taught English classes. One of the teachers with whom I taught would interrupt me often during classes and would attempt to correct me in front of his students. This situation frustrated and annoyed me because, upon first glance, it felt to me as if he was attempting to undermine my credibility in front of the students in an effort to boost his own.

Upon assessing the situation from his perspective however, I could understand that my friendly and engaging style got the kids really motivated and happy to learn. Perhaps he was noticing a positive effect in his classroom that he had never seen before and was afraid that the children would begin to favor me over him. Additionally, this teacher may have felt scared and threatened by the presence of a native English speaker's knowledge in his classroom. Perhaps this was causing him to fear appearing less knowledgeable to his students. And if that happened, he would lose face -- one of the worst things possible in Japanese society.

I initially thought to confront this teacher directly about the issue. To try to talk it out. Just as I would do if it had occurred in The United States. *But it didn't occur in The United States.* It occurred in Japan. So, I decided against that course of action, based upon what I had learned about the non-confrontational nature of the Japanese society and its people.

Instead of addressing the issue directly, I decided to work with him more and on a closer level. I made an effort to see if he needed help with lesson planning. And to support him in any way possible. In the classroom, I began interacting with him more instead of conducting my portion of the lessons separately from his. The results were astounding. Our relationship began to blossom. We were no longer playing for *opposing teams*. We were now playing for the *same team*. And the students could sense it. We worked together to incorporate fun activities into our lessons. We orchestrated role-plays and got the students even more engaged through the use of props. We even set aside time at the beginning of each lesson for us (mainly me) to sing and dance to songs!

Noticing the astounding and marvelous effects of this action, I took the time to plan lessons with other teachers too. And would frequently ask if they needed any help. The vice principal almost always looked busy, so I was usually hesitant to bother him. But I did anyway. A lot. I consistently asked him if there was anything he needed my help with. He would never say "yes," but deep down, I sensed that he appreciated me asking. Through this experience, I learned that work in Japan tends to be more about building relationships and the *process* of getting to the results rather than the actual *results themselves*.

I was used to being given specific work to do and having a clear understanding of the task/s at hand back in The United States. I wasn't given firm direction in that school and therefore was a bit confused at first. I wasn't sure what I should be doing with the free time I found myself with between lessons, so I had to figure out ways to be productive. And do things that would be helpful to my students, the other teachers, and the community as a whole. I couldn't just do nothing with my free time, or at least I couldn't seem like I was doing nothing. I was living in Japan, interacting with Japanese people and with Japanese culture, and therefore had to save face too. This experience taught me the value of proactively seeking out opportunities to help rather than waiting to be tasked with things to do.

So I decided that learning hiragana and katakana (the two Japanese alphabets) and learning kanji (the Japanese characters) would be beneficial because it would allow me to better understand and be able to interact with my coworkers, my students, and my environment. Little did I realize the groundbreaking impact that taking this action would have, how proud I would become of myself, and how much respect I would gain from the community for doing this.

I knew almost no Japanese when I began the role, yet just three months later, I stood up and gave a speech all in Japanese to the faculty and staff. The room went silent when I began. Shock. Disbelief. Many teachers didn't know I had learned how to speak their native language. I thanked everyone for all of their time and hard work and expressed how appreciative I was for having had such an amazing and powerful teaching and learning experience at their school and within their community. I felt my eyes well up with tears as the room erupted into a sea of extended applause.

Through living and working in Japan, I came to understand the value of team building, so when invited out with coworkers, I often accepted. I learned that I had to devote a significant portion of my social life to work-related events in order to maintain the relationships and the harmony that was needed for success and my continued happiness both at work and in that society overall. This experience also taught me how to be more thoughtful and how to build more meaningful relationships.

I participated in and contributed to that community in every way possible, providing genuine praise and plenty of it in every situation that I could. In the classrooms. In the hallways. During extracurricular activities. At social events outside of work. You name it, and I was there dishing out support and encouragement!

I went out socially with teachers and my vice principal.

I helped the kids clean during cleaning time.

I made an effort to support clubs like the judo club by learning about the martial art, going to a practice, and even getting involved in the practice itself.

I watched the baseball team.

I praised the staff in the cafeteria for their hard work.

I attended the graduation ceremony.

I would arrive early to school and watch the band practice. When I first encountered one trumpet player, I saw a timid yet friendly young man. I saw a younger version of myself in him. Having also played the trumpet in my school band, I felt an instantaneous bond with this young boy. I saw an opportunity to be a mentor and took it. He certainly wasn't a professional yet, but I knew that with positive support and attention, this shy boy could develop into an all-star brass player. With my praise and encouragement, that's exactly what he did! He wanted to make me proud, and he practiced very hard.

I showed up to their concert. When the band members spotted me, that boy particularly, their humdrum facial expressions were instantly replaced with looks of surprise and excitement -- they had no idea that I was going to be there. Perhaps a foreign teacher had never come to support them in such an endeavor before. The concert was a great success! And the trumpet player put on a stellar performance! Within just three months, he had developed into one of the finest trumpeters I have ever seen. With a few small acts of kindness and positive reinforcement, I transformed this boy's life. **Sometimes all we need in life is for someone to pay us a compliment, instilling confidence in us. And bringing to our attention a whole new version of, and vision for, ourselves.**

I gave to that community and that community gave back to me. I received massive amounts of appreciation letters and recognition at the end of the school year. My contract got extended. The vice principal of the middle school asked me if there was any way I would be able to stay on and teach with the school again the following year. This was extremely rare because each Japanese public school usually receives a new native English teacher each year. It was clear that within just three months of working at this school, I had made a MAJOR impact on that community. And that community had made a MAJOR impact on me.

Has any event in your life given you a new perspective on your work or caused you to adapt your work style?

Travel also teaches you to improvise and think on your feet.

After working for that Japanese middle school, I worked for a private language learning solutions company. I was asked to teach an advanced English examination preparation course. Recent changes had been enacted in the company, and paid preparation time was no longer being given to teachers to prepare for courses such as this one. This depleted the staff's motivation to prepare for these classes and decreased morale substantially.

I was given a stack of sample tests to hand out to the students during my first class. The idea was to give them a taste of the course, what they would learn in it, and to set their expectations for the rest of the semester. The lack of preparation involved meant that the answer sheets were not double-checked for accuracy. When I delivered the practice examination, I found out that the answer sheets did not match up with the practice tests, rendering the tests useless. This blunder destroyed my plan for the entire first lesson and had the potential to destroy my reputation. And reputation is a hard thing to rebuild. Especially in Japan.

Instead of crumbling in the face of adversity and letting this experience ruin me, I turned a problem into an opportunity. I took all the tests back and said "You see? Life doesn't always go as planned." I then improvised to show them that during the test they were preparing for, and in life, unexpected things can happen. I proceeded to teach a brand new lesson, created on the spot (the students didn't know that), based on this premise. I also used the mishap with the tests to enforce the importance of focusing on what we can control, relating this concept to both the test and to life in general.

Have you ever had to change your plans on the spot due to an unexpected circumstance? What happened and how did it turn out?

Travel also teaches you leadership.

Surely one near-death experience was enough. Life wouldn't put me through another one, right? And even if it did, what would the odds be that it would happen in the SAME country? Too low to worry about, right? Nearly impossible. NEARLY impossible. But not COMPLETELY impossible...

It was a hot day, and I had just crossed the border from Bolivia to Argentina by land. I had booked a bus to go from the border to the Argentine town of Salta. The time came when the bus was supposed to arrive and...no bus arrived. I checked in with the woman at the bus stand who said it was coming. An hour later, she said it was coming. Two hours following that, she said it was coming. Finally, she admitted that the bus indeed was NOT coming. It had broken down, and she didn't know when or even IF it would arrive.

So I searched for transportation alternatives. It was getting late, and the amount of options were diminishing. I could have jumped in a vehicle shared with 15-20 other people, but I preferred not to do that as it seemed it would take quite a long time to fill all of the seats. Nightfall was approaching, and I began to get antsy about getting to my next destination.

On an important side note, I was also battling a raging sunburn that had occurred during my Uyuni Salt Flats experience in Bolivia. My feet and ankles were covered with blisters larger than golf balls due to the irritation caused by the affected areas rubbing against the inner portion of my shoes. No exaggeration. Not. Good. At. All.

I reached the point where I was ready to pay for a taxi. Even those weren't coming. I consulted with police officers, one of whom finally instructed that I, along with a South Korean couple I had just met, go into a black, unmarked van that was headed to the town of Jujuy. At least from there, I could find transportation to my ultimate destination, Salta. But alas, it was an unmarked vehicle, and I should have followed the sage advice that says "don't get into cars with strangers…"

So, I blatantly disregarded that advice and got into a car with strangers. A van, actually. And so did the South Korean couple. But as we started to move, something didn't feel right. The driver and the woman in the front passenger seat didn't seem friendly at all. They were fidgeting. They seemed nervous. The scene became more eerie when the driver asked if we had our passports. I translated this question into English for the South Korean couple, and they told me they did. I then translated and relayed this information in Spanish back to the driver. 'Why would he care about that?' I thought. 'Why would we need our passports? We were not going to be crossing another border…'

And before I could finish that thought, another van with tinted windows emerged out of nowhere, and in a scene which felt like a real-life version of the popular United States television show *Cops*, the van ran us off the road. Two men were in that van. One of them came over to the front passenger side of our van, pulled the woman in that seat out, and took her spot. The other man escorted the woman who had been removed from our van over to join him in the other van -- the van that had pulled us over.

My heart was pounding. I didn't know what was happening. I felt scared. I was nervous. I began thinking, 'Where are we being taken? What do they want from us? Will we be killed? And who are the "bad guys" here?'

Our driver was instructed to follow the other van. After just a few minutes of driving, we arrived at a very small building that had an Argentine flag waving over the main entrance. We were brought inside, and I translated between the South Korean couple and a person who I believed could have been an Argentine Border Patrol Officer. But I wasn't sure. Our bags were thoroughly searched for any type of drugs. Fortunately, neither the couple nor I were in possession of any illegal substances. I never am.

The following was the conversation that ensued between the woman from our van and the officer who was patting her down and questioning her:

> *Officer:* "¿Y qué pasó con tu tío?" (And what happened to your uncle?)
>
> *Woman:* "Se murió." (He died.)
>
> *Officer:* "¿Cómo se murió?" (How did he die?)
>
> *Woman:* [pointing to the corner of the left side of her mouth, and letting her finger slowly slide down her chin while staring intensely at the officer] La coca... (Cocaine...)

I glanced at her feet. The officer was extracting brick after brick. After brick. After brick. Of a white substance that clearly was not sugar from her bags. The same bags that were in the van we had just been riding in.

It was in that moment that I realized what was going on. We were in the middle of a major drug bust. The man and woman had been on their way to Jujuy to deliver cocaine and figured they would make some additional income along the way by taking tourists with them and charging them for the ride.

I gathered my senses and my thoughts and pulled the South Korean couple aside when I found a moment during which everyone else was distracted. The couple didn't speak much English, so I made sure to speak slowly and clearly. I said the following to them "They (pointing to the driver of our van and the woman who had been in the front passenger seat) are transporting illegal drugs. I am going to get out of here. I would recommend you follow me." Without any hesitation whatsoever, they agreed.

Thinking we were unaware of what was happening, one of the officers came over to us and told me that everything was fine. That this had just been a routine passport check. He instructed us to get back into the van with the man and the woman. I refused. He reiterated that everything was fine. I asked the officers to drive us back to the bus station. They refused.

So the South Korean couple and I walked back to the bus station at that late hour of the night. We were unsure exactly where we were going. The blisters on my feet were causing excruciating pain. But we had successfully removed ourselves from the shady operation that we had accidentally and unknowingly involved ourselves in just moments prior. We just wanted to get to Salta. We eventually arrived at the bus station. And just in the nick of time! We caught the one and only (direct) overnight bus of the day. And we arrived in Salta the following morning.

We were in a van with people who were transporting drugs. With that information alone, you might think you know who the "bad guys" are. But after the incident, I learned that it is not uncommon (actually, it is quite common) for Argentine police officers to turn around and sell confiscated drugs for a profit. Were the police officers actually the "bad guys" in this situation? Were both parties the "bad guys"? Is there even a need to label the two parties as "good" and "bad"? This additional information that I gained allowed me to analyze this situation in a new way.

We could have gotten caught up in the middle of a drug deal that went terribly wrong, leading to being kidnapped, shot, and/or killed. I felt grateful and fortunate for having developed the ability to understand and communicate in Spanish -- my language skills may have saved both my and the South Korean couple's lives. I exhibited the leadership necessary to take control of a dangerous situation and led us all to safety.

Have you learned and/or exhibited any leadership qualities through any specific experiences you have had?

CHAPTER 9:

COMMUNICATING ON ANOTHER LEVEL

By traveling, you will absolutely become a better communicator. You will learn how to communicate across cultures, navigate challenging environments, and potentially even learn a new language.

BODY LANGUAGE

Body language differs from culture to culture. For example, when I interact with a Brazilian person, I generally stand closer to them than I would to an American. When I interact with Japanese people, I stand further away. While distance preferences certainly may vary from person to person within the same national culture, the national culture itself tends to play a strong role in dictating these norms. This is just one of many examples of how I tailor my body language in accordance with the culture of the person with whom I am speaking to in order to communicate most effectively.

An interesting phenomenon I observed while in India was something I call "the head bobble." I mentioned this a bit earlier, but I'll go into more detail here. This act consists of wobbling one's head from left to right several times. I was very confused when I first came across this. I wondered what it meant. I noticed that in Southern India, people did this head bobble when it appeared that they were happy.

I found that in certain situations, the head bobble was used when an Indian person was confronted about something they were doing that I perceived to be dishonest. For example, I confronted a driver in India when he asked for a lot more money at the end of the ride than he and I had initially agreed upon when I had entered his vehicle. And I received the head bobble as a response. My interpretation of this body language in that setting was that it was a way of backpedaling. A way of reluctantly admitting that the person had not been approaching the situation in the most ethical manner.

I noticed that sharp movements to either side, a variation of the head bobble, appeared in seemingly heated exchanges, and I perceived that the person engaging in this behavior wanted to get their point across strongly. Overall, I found that the head bobble could mean "yes," "no," or "maybe." Or it could be used to show acknowledgment of an issue, but not necessarily agreement. I learned that this (confusing to me) body language must be examined on a situational basis to best understand its meaning.

Being aware of other people's body language cross-culturally, being aware of our own, and learning the appropriate ways to adapt are some of the key elements involved in successful cross-cultural interactions.

GESTURES

Gestures are another common way that people communicate within and across cultures.

During the world trip I took, I was on a plane flying to Milan, Italy. I was sitting next to a gentleman who only spoke Italian. I named the two pizza restaurants I definitely wanted to visit. When I mentioned one of them, his eyes opened widely. He raised his hand to eye level and proceeded to bring it downward in front of him in a rolling fashion as if he were unraveling an imaginary red carpet in front of us. At the time, I felt it probably meant one of two extremes -- that the restaurant I mentioned was either THE BEST or THE WORST. I later learned from an Italian friend that this gesture indicated that the place we had been speaking of was THE BEST in that man's opinion! And after actually visiting that location, I couldn't disagree with him.

It is very important to understand what gestures mean in specific cultures because they can carry one meaning in one culture and a very different one in another. For example, people often stick up their middle finger, with their palm facing themselves in The United States in an effort to say "F*$% you!" In The United Kingdom however, the upwards extension of the primary and middle fingers on one hand with the palm facing one's own face, is the gesture that portrays this same message. We must therefore be aware of the gestures we are using, and we must be aware of what they mean culturally to the people we are interacting with.

LANGUAGE

Language is an incredibly important communication tool! The value of learning and speaking another language is immense. It allows us to better communicate with and understand people and to build deep and trusting relationships. Fostering and nurturing these relationships carries with it the potential and *extremely likely* possibility of bringing us more friendship, romantic, and professional opportunities.

When I arrived in China, I spoke no Mandarin Chinese. Shortly after arriving however, I realized that very few people spoke English. Because I was planning on spending more than a month there, I saw the need to learn at least a bit of Mandarin to be able to communicate with Chinese people in their native language. That is my rule -- if I'm going to spend more than a month in a place, I am going to make a concerted effort to learn how to say more than just "hi," and "how are you?" in the location's primary language.

I began to learn the most basic phrases. Naturally, at first, I heard phrases like "Wo ting bu dong" (I don't understand) and "Wo bu ji dao" (I don't know) because people were not understanding me. Then, I began to learn basic words like "jigga" (this), and expressions like "wo yao" (I want). With these phrases, among a few others at my disposal, I was able to point to items on a menu and tell the person taking my order what I wanted. This simple gesture of making attempts to speak with Chinese people in their native language went a LONG way. **IT COMMANDED AN ASTRONOMICAL LEVEL OF RESPECT.** Not only were the people I came into contact with astonished in the best of ways, they were delighted. I imagine this could be because most foreigners who the Chinese people I met had come into contact with, if any, had not made this kind of effort to communicate with them in their mother tongue.

As my travels throughout China continued, I began to learn more Mandarin Chinese. I received kindness on so many levels, often having people offer to pay for my food and drinks. On one occasion, someone I sat next to on a bus refused to let me pay for my own hotel room that night! And that was not the only occasion when someone else paid for my hotel room in that country. The generosity I experienced was, in my mind, a combination of hospitality and kindness on their part, fueled by my curiosity to learn more and my burning desire to better understand and communicate with others.

One reason why speaking to people in their native tongue is so powerful is because PEOPLE CAN SENSE THAT YOU ARE CURIOUS ABOUT THEM AND THAT YOU RESPECT THEIR CULTURE. IN TURN, THEY GAIN RESPECT FOR YOU! They will likely feel that you care enough to want to actually know about them. They will realize that you are willing to step outside of your comfort zone, be vulnerable, risk making mistakes and sounding silly in a language that is not your first, all in an effort to learn about and understand them better. And in turn, they will likely become more curious about you and your culture too!

Although more challenging for you or me, speaking, reading, writing, etc. in another person's first language facilitates communication. It eliminates the need for the other person to worry about making mistakes in English (or another language) and therefore puts them at ease. Their words will flow freely because their mind isn't working to translate their ideas from one language to another. And the ideas they express are more likely to be genuine. And to come from the heart.

The biggest risk of communicating with someone else in their native language (one that is not yours) is that you or I don't fully understand the meaning a person is trying to convey due to our lack of vocabulary, grammar, etc. in that language. In which case, we can ask the person to slow down and to repeat the word/s or phrase uttered. And if after trying that we are still struggling to understand, we can use a translator to capture the meaning.

Learning a new language also allows for new levels of understanding to be unlocked because true and heartfelt messages are being transmitted. Which allows for deeper connections to be made.

During my MBA, I was invited to stay with my classmate's parents in Rio de Janeiro, Brazil. His mother and father speak no English. I had a choice. I could have just spoken to them in English, hoping they would understand at least a tiny portion of what I was saying. Or, I could have tried to make a solid effort to communicate with them in their native language, Portuguese. I chose the latter. I had begun participating in an online language exchange before my first trip to Brazil and ultimately ended up meeting two people, in person, in Brazil from it! However, it was the experience of staying with my friend's family that catapulted my Portuguese skills to new heights. I began to understand more about his mother and father. I began to see them as parents. I began to treat them as parents. My parents. They began to see me as a son. They began to treat me as a son. Their son.

When I first arrived at their place, my friend's father and mother greeted me with big hugs. They brought me inside, and the father said, "Tira ropa." (Take off your clothes.) Due in part to his accent, and because my Portuguese wasn't so strong yet, I couldn't understand what he was saying. He then repeated this phrase two more times, and with more zeal and energy, "Tira ropa! Tira ropa!" (Take off your clothes! Take off your clothes!) he exclaimed. And as my friend's father grabbed the bottom of my shirt and began lifting it up, my brain made the connection. I thought to myself, 'Okay, I think I understand what he's saying. He wants me to take my clothes off. ...But why would he want me to do that? Why would a man I just met want me to be sitting in his house naked? .. ..'

And within five minutes of meeting my friend's family, I was in nothing but my underwear. His father had won the battle. He had gotten what he wanted. And he was content. I was not. I was uncomfortable. I was borderline nude. I felt embarrassed. Inappropriately dressed, sitting scantily clad on this family's couch. I was unable to understand why this had happened. And then his father placed a container of freshly baked goods on my head.

Upon speaking with my friend later on, I learned that his father asked me to remove my clothing because he wanted me to be comfortable in his house. He knew it was scorching hot outside, and he thought that me taking my clothes off would be the best solution to the problem.

The container of sweets placed upon my head had nothing to do with the heat. That was due exclusively to his quirky personality. I would later learn just how similar he and I truly are. I would come to appreciate finding someone in life who is just as silly as I am. In another culture. Who speaks another language. Had I never made the effort to learn Portuguese, I never would have realized that. And I never would have connected with him on that level.

The father of this family is quite bold, however he is one of the most amazing and well-intentioned people I have ever met in my life. He has worked so hard, done so well for himself, and has given so much away -- love, money, possessions, etc. -- just to find himself with even more of an abundance of all of these things. He has further reinforced in me the value of education, working hard, being selfless and helpful, and putting others' needs and interests before my own.

I have learned the value of joy, kindness, and giving from his wife. She has gone above and beyond to make me feel comfortable and to support me in every way imaginable and more. Making sure I had food to eat, a comfortable place to sleep, clean clothing, and so on. During this visit. And during every subsequent one as well.

This family treats me like a son. Because, in their mind, and in mine as well, I AM their son. They bring me to their neighbors' homes to visit the elderly and the ill or just to engage in conversation. They bring me to barbeques. They bring me to events. They ask me to show their friends around town. They brag about my accomplishments just as they do about their own son's. And I assure you, his accomplishments are far more impressive than mine. I have become a part of their family. And they have become a part of mine. And for that, I am and will be forever grateful.

It was my desire to better understand this family and the other amazing Brazilian people that I met that fueled my inner fire to be able to communicate in Portuguese. And the Brazilian people reciprocated with love and open arms. They were so happy that I was making the effort to learn Portuguese and that I was able to communicate with them in their primary language! And they still are. Had I never learned Portuguese, I never would have had the amazing ability to understand and be impacted in such a powerful way by this amazing family or by the other wonderful Brazilian (and Portuguese) souls I have met over the years and throughout my travels. I keep in touch regularly with this family as I consider them my own.

Learning and utilizing another language is also extremely beneficial on a professional level.

I was a Spanish major in college and then studied abroad in Argentina. I felt that I had a solid command of the language... until I arrived in Buenos Aires and realized I understood practically

NONE of what was going on around me. The porteños (people from the city of Buenos Aires) have quite the distinct accent, and combined with all of the slang terms used in normal conversation, I had a lot of difficulty understanding the language. I had studied Spanish for years, but this was the first time I had been immersed in another place where the first language spoken wasn't English. And where I needed to be consistently communicating in Spanish throughout the day. Each day. Every day. I found that using the language in a real-world setting was very different from communicating with classmates using textbook exercises. As I continued my efforts however, and continued to practice with classmates, teachers, and my host family, I ultimately arrived at a level of fluency! And oh how proud I am of this feat!

As I gained a better understanding of crucial elements of Argentine culture (dance, music, food, art, etc.), I was able to connect on a deeper level with Argentine people, allowing me to create more meaningful and long-lasting relationships.

One such relationship that stands out in my mind is the bond I created with my final host family, The Arnaudos. The mother and father of this family are actually the co-owners of the company that had arranged my first and second host families. They were aware of the scary incident that had occurred in my first host family's home and knew that I wasn't overly happy in my second. But since they had two little girls, they had only ever housed female exchange students in their own home. I could understand the concerns associated with having a male exchange student living in their apartment. And could therefore see why they had never allowed it.

I was surprised that the father of this family, Ignacio, would invite me out to do things. Pretty frequently. I must admit that I was skeptical at first. 'What does he want?' I thought. And as time went on, I learned EXACTLY what he wanted. He wanted to be my friend. To show me around his fabulous city. To have me bond with his wife. And his loving children. And his adorable dog.

As a Catholic, Ignacio wanted to take me to a church service. Not to convert me. But to invite me into his world. To open up my mind to his way of life. A way of life unbeknownst to me prior. There was no pressure. Only a sense of exploration and learning. So I accompanied him. I learned a lot. And I enjoyed the experience. Very much. And he sensed it.

He had always been curious about Judaism, and after attending that church service together, he told me he wanted to experience a service in a synagogue. So one weekend, off we went together to attend a service in a synagogue. He really enjoyed the experience. So did I. We later discussed the similarities between the two services we had attended together. And we came across a rather striking revelation -- *perhaps the world's religions aren't so different after all...*

Ignacio became my first true friend in Buenos Aires. We both enjoyed learning from one another. Exchanging exciting ideas and having interesting discussions. And having a ton of laughs all along the way! We would go out for late afternoon coffees together. We would go to his favorite pizza restaurants. I would go to his home for dinner. He took me to his favorite soccer team's stadium, La Bombonera. He and his family would take me out of Buenos Aires and to the city of Tigre to row. And Ignacio and I became, as he so nobly deemed us, "The Lords of The River."

I went on to become the first *and only* male exchange student to live with that family. Because I had built friendship. Because I had built rapport. Because I had built trust. And because I had become the son they never had.

I will **NEVER** forget how Ignacio and his family made me feel. While so far from my entire support network of friends and family, this family made me feel at home. This family made me feel welcome in a country so far from my own. Most importantly, this family made me feel **loved**. *This family* became *my family*.

And it should come as no surprise that this family and I are still in contact today. I have since been back to Argentina to visit. My host father and I were even afforded the amazing opportunity to travel throughout Argentina together by car for over a week! We experienced some of the country's most beautiful nature and saw some of its most incredible wildlife. What a wonderful bonding experience that was, and I am forever grateful that we were both able to take the time to do that. Bonding experiences such as those, in my eyes, are both necessary and important to have, if and whenever possible.

I was so delighted to finally have the opportunity to show my Argentine family around when they visited New York City! I took them out for the best bagels, phenomenal pizza, and to see attractions such as The High Line! I wanted to ensure they had the best experience possible in my city, just as they had done for me years prior in theirs.

In between my run-in with a massive amount of illegal drugs on the border of Argentina and Bolivia, and before arriving to Buenos Aires to reunite with Ignacio and his family, I made a stop in Córdoba. I had heard that when the people from this city speak, their tones inflect up and down in a way that sounds to many like they are singing. But that wasn't why I was there. I was there because my aunt had studied abroad there years ago. And she had asked me for a favor. She had asked me to visit her host family while in Argentina.

Does that sound ridiculous to you? To visit a family you've never met before?

Well, it sounded ridiculously AWESOME to me! So I told my aunt I would gladly go.

When I arrived at this family's home, an instant connection was created. My aunt's host mother, host sister, and others greeted me with hugs, kisses, and an indescribable aura of warmth. They were curious about me, and I was curious about them. They talked while I listened and learned. Then I spoke while they listened and learned.

I was home. I was spending time with a family I had never met before. I was in a house I had never been to before. I was in a city I had never visited before. Yet *I was home.*

What amazing and wonderful connections we can make and experiences we can have in this world. Across borders. Across languages. Across cultures. If we just allow ourselves to be open to connecting with others, take the time to build meaningful relationships, and let life take its course.

After relocating back to The United States from Japan, I took on a role with a company where I arranged cross-cultural training programs for their clients. In that role, I arranged training sessions for employees relocating from one office to another. Usually internationally. Sometimes domestically. I managed the communication between the cross-cultural trainer and the relocating employee to ensure the smooth delivery of the training.

Most trainings were in English, but we received some requests for training sessions that were to be delivered in Spanish. Because of my strong Spanish abilities (speaking, understanding, listening, writing, etc.), I was asked to handle the programs where people relocating had a first language of Spanish. This meant translating email templates from English to Spanish, reading and responding to emails in Spanish, speaking Spanish over the phone, etc. I gladly accepted this responsibility and soon became the point person in our office for handling Spanish-speaking programs. This boosted my resume and qualifications. More importantly, this boosted my value in the workplace. And *even more* importantly, it boosted my confidence, allowing me to impact and serve others on a greater level. It also served as an excellent way to maintain and improve my Spanish language skills. From this experience, I learned the value that knowing an additional language carries and of the professional opportunities that can follow suit because of it.

You know why you shouldn't feel afraid to learn or speak another language? Because **WE DON'T EVEN SPEAK OUR OWN FIRST LANGUAGE PERFECTLY! SO WHO CARES IF YOU SCREW A FEW THINGS UP IN ANOTHER LANGUAGE?!** I guarantee you that *you* care a whole lot more than anyone else you will communicate with does. Actually, they will care, but care in the sense that they will be so amazed at your courage. Mistakes are how we learn, and without them, we'll stay stagnant and never grow. So don't be afraid to get out there, screw up a few words and/or phrases in a new language, and make some deep connections by communicating with people from other places and cultures! If I can overcome the fear of making mistakes to improve my foreign language skills, so can you!

When we understand the body language, gestures, and language of a new culture, we come upon a powerful realization. We learn that *we* have the power to impact the communication. We have the ability to empower ourselves and others by tweaking the three aforementioned elements -- body language, gestures, and language - in order to communicate more effectively.

One of the biggest challenges you may face, especially if you relocate to a new place, is striking the balance between being who you are (or who you believe yourself to be) while adapting your personality (at least to some degree) to the culture of the new location you are in. You'll want to find the happy medium between acting how you are accustomed to acting and acting the way one is expected to act in the new place. Understanding and making the necessary adjustments. And to the right degrees. Being 100% your old self most likely won't work. Neither will attempting to replicate EXACTLY what the locals do as that can be perceived as "fake" or "trying too hard to fit in." Meeting somewhere in the middle with your original and new personalities and communication styles is the key to cross-cultural integration -- it isn't easy, but it is vital for success.

In Japan, I realized that the "American" -- individualistic, bold, and opinionated Rob won me no friends and only strained my relationships. Bowing and smiling all the time, however, felt fake and disingenuous. I had to find the balance between the two cultures to arrive at a hybrid personality and communication style that worked and was authentic while in Japan. And I did. It wasn't easy, but that is something I am very proud of having accomplished. Furthermore, as a result of over 15 years of international experience and exposure to diverse people, communities, and cultures worldwide, I have developed a keen ability to culturally "fit in" or "adapt" instantaneously to new environments. *A cultural chameleon if you will.*

Have you had a similar experience? Have you had to adjust your body language, gestures, and/or learn a new language to improve communication in either a personal or professional setting? If so, what was the situation, and what did you notice when you made adjustments to your communication style? Were the interactions or relationships impacted in any way? If so, how?

CHAPTER 10:

DETERMINATION AND PROBLEM SOLVING

While traveling, you either have or will come across problems that must be solved. Doing so will inevitably make you a better problem solver. The challenges you will face will usually fall into three main categories: physical, mental, and spiritual. I have found it both helpful and important to remember that regardless of which category the challenge falls into, we do not have to *impulsively react*. **WE** have the ability to **choose our response**.

PHYSICAL

At some point during your trip, you will likely encounter a physical challenge or two. Or seven. Or 147. This can come in the form of a hiking trail that you find more difficult than you had originally anticipated. Or perhaps you injure yourself somehow. Either way, it's important to assess the situation with care and to move along with your journey appropriately, using the experience as a catalyst for growth rather than letting it cripple you. While never fun at the time, encountering these issues will invariably turn you into a stronger and more resilient person. A more resourceful traveler. And a better problem solver.

During my most recent world trip, my body was challenged in a way I had never experienced before.

I had traveled through a good portion of The Caribbean, and France was my next stop. With a layover in Miami. When I arrived at the check-in counter in Miami for my flight to Paris, I was told I could not enter the plane because I did not have an onward ticket (leaving France). Lesson to be learned here -- you may be asked for proof of return or onward travel when traveling internationally. Without said proof, you are likely not getting on that airplane. Although I was annoyed at the time, I had other issues to worry about. Like the illness that had been attacking my body relentlessly for the past two weeks. I had been feeling quite ill because of it. And I didn't seem to be getting any better. So perhaps not being allowed on that flight was a sign from the universe...

So I opted against purchasing a last-minute onward ticket and decided to stay in Miami to figure the situation out. It was good to be in The United States, with access to an advanced medical system. I went to an urgent care clinic, and even through tests, was unable to identify what the issue was. I learned that my white blood cell count (created by the body to fight off infections) was elevated.

Actually, it was more than double the normal amount. This was VERY concerning to the doctors. This was VERY concerning to my family. This was VERY concerning to my friends. This was VERY concerning to me.

My body continued to fight, day and night. I would wake up severely dehydrated and with soaked bedsheets due to intense night sweats. I would walk downstairs to the lobby of the hotel, hang out there for a few minutes, but would be unable to stay there too long as I felt chilled by the air conditioning. I would grab some vegetarian sushi or veggie burgers and drink lots of iced green tea with the aim of improving my situation. But I still didn't appear to be getting any better. As much as I tried to eat healthily and do my body good, it didn't seem to be working. Nothing I tried seemed to be working.

I was told that I would need to keep an eye on my white blood cell count and that if it stayed as high as it was, it would be advisable for me to cut the trip short and return home. I continued getting tested periodically -- for three weeks -- to monitor my white blood cell count. If it didn't begin decreasing, or if it began increasing, I would need to see an infectious disease specialist.

This situation was extremely stressful, however through it, I learned how to be resourceful and how to effectively handle stress while working through the uncertainties of my medical condition. I felt confused. Scared. And unsure what to do. 'How long should I wait for the white blood cell count to descend? How often should I check in with urgent care? Should I go back to New York? Should I just stick it out here in Miami? Will I be able to continue my trip, or will this be the end of my travels?' These were the thoughts that were running through my mind.

I could have quit. Cut the trip short. And returned home to my mother's house. She would have welcomed me with open arms and taken care of me. Good care of me. Excellent care of me. World-class care of me.

But I didn't quit. I didn't fly home. I kept fighting. And *fighting*. And **fighting**. And **FIGHTING.**

Although I did not have physical access to our family friend who is a doctor or to my friend who is a pharmacist, I knew I could get a hold of them over the phone and via email. So I did. And I gained information. And helpful opinions. And recommendations on how to proceed. Along with much needed emotional support. I also stayed in touch with my family, providing updates and receiving additional ideas, guidance, and support.

I befriended the hotel staff, informed them of my situation, and they generously offered me plentiful amounts of water bottles to help keep my body hydrated. I ended up being the second-longest staying guest that the hotel had ever had. I appreciated the hotel and it appreciated me. And so the story goes that the hotel and I fell deeply in love and got married. Except that last part didn't actually happen.

My white blood cell count eventually began to go down and decreased enough for the doctor and our family friend in the medical field to agree that I would be alright to continue traveling again.

As it turned out, I had picked up a virus. A very strong virus. I don't know from where. I don't know how. And I don't even know what it was.

But what I DO know is this: I overcame the situation. I won the battle. Through patience. Through persistence. Through advice from trusted sources. Through difficult decision-making. Through positive thinking. Through resilience. Through willpower. Through the synergy of my mind, body, and spirit firing together on all cylinders. I was determined to continue traveling, and the universe seemed to respond accordingly, respecting my intense resolve and allowing my body to overcome this seemingly debilitating illness in order to continue my journey.

On this trip, my strategy was booking one location at a time. It accounted for the potential of something like this occurring and worked to my advantage in this instance. I only had to cancel one flight and didn't have to cancel any hotel bookings. In fact, I had purchased travel insurance prior to my trip which ended up covering a good portion of the hotel and some additional costs pertaining to my illness while stuck in Miami. I had won on a financial level too.

Experiencing so much uncertainty and my own vulnerability during this medically challenging situation was very humbling. Working through it and overcoming it was transformative.

If I can overcome physical challenges on my travels, so can you.

Have you experienced a major physical challenge in your life? If so, what happened, and what was the result of it?

MENTAL

The illness I was confronted with on that trip most certainly required physical strength, however mental strength was required to overcome it as well. In this section, we'll focus on some of the mental challenges you may go through that will strengthen your character and help you to become a better problem solver.

I am vegetarian, and during a trip to Bolivia, I was invited to a friend's family's home for lunch. And chicken was served. Her mother knew that I was vegetarian but told me I didn't have to eat the chicken. I was in a tough situation because I wanted to be respectful of Bolivian culture and of my friend's home and family. However, I also wanted to respect my beliefs and stay in line with my morals and values. As I ate, I saved the piece of chicken for last. My friend's mother looked at me and told me again (in Spanish) that I didn't have to eat it. Part of me felt bad, but I didn't end up eating it. Although I didn't consume the food, I felt that I had shown respect by considering eating it and not refusing it outright.

By allowing ourselves to have novel (often cross-cultural) interactions, we are able to come up with creative solutions to issues we never even realized we might encounter. And we can learn to think on our feet and to handle them with tact and poise.

In another instance, while on a moving train in India, I picked an empty wrapper up off the train floor and began searching for a place to deposit it. An Indian man took note of my predicament and snatched the wrapper away from me. My belief was that he knew where a garbage can was and would dispose of it. He did. And he did. Yet to my surprise and bewilderment, instead of throwing the piece of garbage into a trash bin, he simply flung it effortlessly out of the moving train, using nature as his garbage can. He then turned to me, and with a gigantic smile growing on his face, said the following, "This is India."

I stared at the man briefly in disbelief as I tried to process what had just happened. I had grown up learning that littering was wrong, and this action therefore went against my moral grain. The more I thought about what he had said -- "This is India" – combined with the heaps of garbage I had seen in so many locations in this country, it started to occur to me that depositing garbage in places that aren't trash cans might not be seen as a bad thing in that society; it actually appeared to be quite a common and normal occurrence. This situation, along with many others -- in India and in other countries -- stretched my mind in new ways and allowed me to view situations from new perspectives.

Your mind is both your best friend and your biggest enemy. In Chapter 2, I discussed fear and how paralyzing it can be. Fear can add immense amounts of emotional stress. Additionally, negative thinking will only fuel this fire, so we've got to make sure to get/keep ourselves out of this zone. That's easier said than done. **This book is aimed at helping you get it both SAID AND DONE.**

Fear can fuel determination and can positively affect how you solve problems.

Before deciding to go to Europe on my own, after learning that my friend would be unable to join me, I felt fear. But I chose to go anyway. *Alone.* Hence, instilling in me the confidence to embark on further solo travels.

Before letting my finger off the mouse to confirm the purchase of that one-way flight to China to begin my Outside-Of-The-Classroom Learning Experience in Asia, I felt fear. But I let go of the left-click button and embarked on what would become my path of passion.

When robbers entered my host family's apartment in Argentina, I felt fear. I could have ended my stay and flown home immediately, but I chose to remain; I could have stopped traveling altogether, but instead, I chose to use this experience to enhance my determination to follow my passion for travel.

When I realized I was in the middle of a drug transportation operation on the border of Bolivia and Argentina, I felt fear. I followed my gut feeling, along with cues from my environment to solve a serious problem, becoming a leader in the process by encouraging others to follow me to safety.

Do you recognize a common theme here? It's FEAR. It's proof that I am human. Fear is something we all experience. On an evolutionary level, fear has served, and continues to serve as a survival mechanism that alerts us of perceived dangers and threats in our environment. But we also experience fear at times when we are not in physical danger, and our survival is not at stake; it seems that the majority of times we experience fear, it is in *this* context. **It is important to recognize fear, but what is MORE important is HOW WE RESPOND to fear.** Responding to fear in a positive and productive manner allows us to overcome challenges and can be both enlightening and growth-producing.

What you'll also notice in the above examples is that **I was able to OVERCOME** the fears I faced instead of letting them dismantle me. Re-read the previous sentence. It's THAT important. If I can transform my fear into motivation to grow rather than letting it stifle me, so can you.

If you don't recall the details of the stories referenced above, you may want to go back and re-read them to remind yourself that TRULY ANYTHING IS ACHIEVABLE.

Can you identify some of the mental challenges you might encounter during travel?

Has fear ever stood in your way? If so, how did it affect your resolve and the way you solved the problem?

SPIRITUAL

There is a larger, often overlooked element of travel that is crucial. I am speaking here of the spiritual realm. Spirituality is a broad concept that is very individual and can vary significantly from person to person. Having our own spiritual experiences amidst our travels affords us the opportunity to better assess our character, our values, and our connection to the universe.

I used to say that I was a "see-it-to-believe-it" kind of guy. Meaning that if I couldn't physically touch or feel it, and therefore had no physical evidence of its existence, I didn't believe in it. I was therefore skeptical about the idea that there could be any higher forces in play in life.

Through traveling and learning about new religions, new concepts, and new ideas, such as Buddhism and karma, my thinking began to shift. I have come to believe in karma as I sense that I have experienced it to be true in my life. However, karma is not something tangible. For that reason, I began to believe that perhaps there could be some greater forces at work, doing things "behind the scenes." I can no longer claim to be a "see-it-to-believe-it" kind of guy. This was a major shift in my beliefs, and I credit travel as the transformational mechanism that facilitated it.

I have often challenged the notions of fate and free will. How they collide. If only one can be real. Or if perhaps they can coexist. Do our choices have an impact on our future? Was I destined to take that Asia trip? Was it 100% my choice? Was it perhaps a combination of the two? What do you think?

I can't claim to have all the answers, but I can solemnly say that traveling and learning new ways of doing things has caused me to question my own set of beliefs. It has opened up my mind to a whole new world of possibilities. A restaurant menu is a phenomenal analogy to use to illustrate this point. Before having traveled, I only had one or two items to choose from. But as I traveled, learned more about the world, and discovered new ways of thinking and accomplishing goals, that menu began to grow. Gradually. Then substantially. Then exponentially. And now, the amount of choices on the menu is overwhelming. But I can't claim to be upset about that.

By constantly learning, changing, and growing through travel, I have discovered new things about myself and have realized new paths I want to take moving forward. My travel experiences have challenged elements of my faith. They have made me think through what I believe in and why. And they have allowed me to connect deeper with my spirit and how it vibrates with the universe.

I have found my spirit to be more connected to the universe. I believe this has been accomplished through the wide spectrum of experiences I have been fortunate to have had through travel. Like the 10-day meditation course in Jaipur, where I connected with myself and with nature. Like the interaction with that woman in the tent in Litang where I connected with another human being. Like my face-to-face gorilla encounter in the wild in Uganda where I connected with another species. The cumulative effect of all of these experiences is that I have felt more connected with the universe. And all of its different components.

This phenomenon has shown me that I am constantly evolving. That I am constantly changing. That I am constantly growing. Taking in and considering new ideas and perspectives as I encounter them. Synthesizing and analyzing new information as I come across it. This process challenges us to embrace change or cling to our former selves. I have found embracing change to be transformative.

What do you believe in? Why do you believe in it? Do you believe in those things due to your upbringing and/or your surroundings? Have you ever had an experience or learned something that challenged those beliefs? If so, how did you react?

Have you had any experiences that opened up your mind to new perspectives? If so, what was it/were they, and how were you affected? Did you make any changes in your life because of these experiences?

CHAPTER II:

VALUES, VIEWPOINTS, AND APPRECIATION

In previous chapters, I discussed the impact that exposure to new cultures has had on both my thoughts and my actions. It has transformed how I see the world. How I understand and conduct business. What I eat. How I approach and maintain relationships. And so much more.

If I can transform through travel and through my exposure to different cultures, so can you.

Engaging with new cultures will cause you to become a more empathetic person. **By exposing yourself to a new society and the culture that permeates it, you will be better able to understand how people think, and more importantly, WHY they think the way they do.**

During my time living in Japan, I entered a shoe store one afternoon. I saw a store clerk down on one knee and a customer standing. The odd thing however, was that the woman on her knees wasn't helping the customer to try on the shoes, as I suspected would be the case. Instead, she happened to be taking a credit card. Let me repeat that. Instead, she happened to be *taking a credit card.*

I couldn't understand, at first, why she was on one knee taking the card. Throughout my time living in Japan and looking back at that experience, I was able to better understand why this had happened. The store clerk was exhibiting respect for her customer, at a level I believe to be unbeknownst to most of the world outside of Japan (and those people who have visited or spent time in that country). She was showing appreciation for the woman making a purchase. She was showing her appreciation for the woman being a customer. Upon better understanding the situation and analyzing it, I found this to be an example of an extraordinarily high level of customer service.

And the above example is only one of hundreds that I experienced while living in Japan. Throughout Japan, you too will likely find the highest level of customer service you have ever experienced in your life. I believe that this is, at least in part, because making others feel comfortable and respected is EXTREMELY important in Japanese culture. Because of how you are greeted and treated, you will probably feel like royalty walking into *any* restaurant. Even a fast-food joint. No need to go to a fancy restaurant to experience a high level of service in that country. **After visiting over 50 countries, I can say with confidence that Japan is the country in which I have received the highest level of customer service. In all interactions. In all settings. And on a consistent basis.**

Another factor to consider concerning the dining experience in Japan is that tipping does not occur. By removing it from the equation entirely, a situation is created where people feel comfortable paying the set price. Nothing less. Nothing more. Customers don't have to feel pressured to leave additional money as a tip. And the servers can do their job without worrying about their performance affecting the amount of money they receive. A win-win for all involved!

Having been raised in The United States, I was accustomed to being involved in interactions where tipping occurs often and is expected in most eat-in dining and/or drinking situations. Having the opportunity to experience Japan's system allowed me to see this phenomenon from a new angle and to process the pros and cons of tipping.

As mentioned prior, I was afforded the opportunity to stay with my good friend and MBA classmate's family in Brazil. Not only did this experience motivate me to learn Portuguese, but it also provided me with a deep insight into Brazilian culture. It became clear to me, through the interactions I had with my friend's family and with other Brazilian people outside of their home, how important relationships are in that country. The simple act of stopping to have a conversation is appreciated. The focus in Brazil tends to be more on relationships than on tasks. The extra steps taken to build and maintain relationships are seen as very important in Brazilian society.

Another interesting occurrence was what I was told by a man while waiting for a bus in India. The bus was late, and I asked him when he thought it would be arriving. "If God wants," he replied. My first reaction was that his response sounded crazy. So *he* was probably crazy. After having more experiences in India however, and better understanding the mythical and spiritual elements that pervade that society, I began to see his comment from a new lens. I began to see the faith behind it.

There is great value in understanding new cultures and their ways of interacting with the world.

Meeting people abroad, I have noticed that most people complain about their government and about their leaders. This seems to be a universal phenomenon. Perhaps all complaints are justified. But perhaps we take for granted the good things these leaders do. What I have taken away from these conversations is that many people tend to focus on the negatives of politics. The egregious things the government is doing. The things the government is not doing but should be doing. And so on. And the negativity seems not only to be tied to politics but to life in general.

Something I have found particularly interesting is being able to see politics from a new perspective. Almost every time I travel outside of The United States, I am asked what I think of our president at the time, of our political system, or both. And I am invariably provided with the views that the person I am speaking with has about the president of The United States, The United States political system, or about United States politics in general. While views differ, one thing remains constant -- EVERYONE has *some opinion* on these things. This has helped me to better understand the powerful role that The United States plays in shaping the world. I have found that whether for the better or the worse, most countries view The United States as a major world power, at least at present.

And perhaps for that reason, most people worldwide follow United States current events, at least to some degree. That is powerful. Not every country has this privilege.

Hearing people from other countries' views about US politics has given me the opportunity to rethink my own beliefs. And I believe that with privilege comes responsibility. I have realized the great responsibility that I carry as a United States citizen each and every time I step off an airplane and enter a new country. I may encounter a person who knows something about The United States, be it as simple as having watched a few American movies or someone who is more up to date with United States current events than I am. To others, I may be the first American they have ever met in their lives. To all of them, I am a "social ambassador" of The United States. To them, I represent The United States of America. That's a powerful sentence; please re-read it. In this context, one person can, and often does, represent an entire nation. And that is a potent impact that one person can have on the world.

Whatever country you are from, please be aware that whether you realize it or not, you are representing your country when you travel abroad. Your words, your behavior, and your actions will shape people's views of your country and your culture. Your words, your behavior, and your actions will reaffirm or contradict their beliefs about your country and your culture. Your words, your behavior, and your actions will serve as an education for people around the world about your country and your culture. **I have therefore learned that it is imperative that I represent and promote a positive image of my country.**

Many folks I have met while traveling have had trouble wrapping their head around the fact that I am American. This may be because I don't fit the stereotypes that some folks around the world hold to be true about Americans. An American who owns a passport? An

American who's traveled extensively? An American who speaks multiple languages? Here is how a typical exchange often goes:

Person Rob is Meeting: "Wait a minute, you're American?"

Rob: "Yes."

Person Rob is Meeting: [With a puzzled look on their face] "… But you…"

And within the blink of an eye, a MAJOR change has just taken place. As the communication progresses and the initial wave of confusion begins to subside, a WHOLE NEW perception of what it means to be American begins to brew in that person's mind. The reset button has just been pushed. A vital paradigm shift has just occurred. The understanding that stereotypes about Americans don't always hold true has just been gained.

The spotlight has been shifted away from these stereotypes and onto the freedom and amazing benefits that American Citizens have and enjoy. Those benefits that allow Americans to travel to nations far, far away. Those benefits that allow me, as an American, to travel to nations far, far away.

The spotlight also gets shone on the individualistic qualities that The United States exudes -- that I have the amazing opportunity to be myself. That I have the amazing opportunity to practice my beliefs freely. That I have the amazing opportunity to shape my own destiny.

And suddenly, thoughts of The American Dream begin to enter their mind as they realize that they are indeed speaking with someone who is LIVING it. RIGHT THEN. RIGHT THERE. AND RIGHT IN FRONT OF THEIR VERY OWN EYES. Someone who has decided to travel to their land for the opportunity to learn ABOUT them. And to learn FROM them.

To me, The American Dream isn't about owning a grandiose home on a substantially large amount of property surrounded by a white picket fence and an expensive car in the driveway. Instead, I believe that The American Dream is about having and utilizing the freedom to pursue my dreams and passions. The freedom to travel to places far away and *to make a positive impact on* the people I meet. The freedom to travel to lands far away and *to be impacted by* the people I meet. **And most importantly, THE FREEDOM TO MAKE DREAMS REALITY BY TAKING ACTION!**

I am thankful for all those throughout history who have fought so hard to gain and maintain those freedoms for me. Just as each of those people fought for freedom in their own ways, I am fighting to show people around the world that The United States of America is a great country! A land of opportunity. **This country is certainly not perfect; however I have been inspired through my travels NOT to sit around and wait for others to change the world, rather to CHANGE THE WORLD MYSELF. If I can change the world, so can you.**

We often tend to overlook, undervalue, and take our freedoms for granted. By traveling the world, I have been able to reflect on my life, and I appreciate how lucky I am for the many benefits and privileges I enjoy as a United States citizen. I don't always agree with political decisions or certain things pertaining to United States society, but travel has opened up my eyes to how much we, as United States citizens, have. And to how much opportunity there is here in The United States of America. To summarize, travel has allowed me to become a more proud and grateful American.

In fact, living abroad inspired me to come back and see the Statue of Liberty, visit the Empire State Building, and partake in other activities in New York City that I previously had access to yet never did. By living overseas, I was afforded a new perspective. I realized how easy it is to take a place for granted and never explore it when we easily have the opportunity to do so. I began to understand why so many people *never* visit these amazing places nor take part in these amazing experiences that their own country has to offer. They don't prioritize them, thinking they have forever to do them. *When in reality, they don't.* With this idea in mind, and fueled by the fact that when I initially arrived in Japan for work I was unsure how long I would stay, I did and saw as much as possible while there.

Can you see how your morals and values could be challenged and influenced by information gained and experiences had through travel and interacting with new people from new countries and new cultures?

CHAPTER 12:

PASSION AND PURPOSE

None of us have a crystal ball, and we therefore don't know what will happen tomorrow, what situation could change, or when our last day on Earth might be. That's why I chose and continue to choose to take on these immense challenges through extensive travel. I therefore urge you to take action TODAY, even if it's just beginning to plan your dream trip. If you can't go immediately, you can begin thinking about and roughly planning, researching, and outlining your trip. You can put as many relevant details as possible down on paper for the time being. And you can always make changes and updates later. The point is to AVOID PROCRASTINATION, TAKE ACTION, and START NOW!

TAKE ACTION! Don't put the things you want to do in life off. Remember that if we keep putting these things off, they will eventually be put off forever and will ultimately never get accomplished.

Travel is an amazing way to re-evaluate your values, beliefs, and morals.

Can you see how travel could allow you to expand your thinking?

One thing is for sure. Travel is POWERFUL. It has helped me to identify and more clearly understand my passions:

- For travel itself
- For communicating in diverse languages
- For learning about other cultures
- For better understanding people
- And for making genuine connections

IF I CAN DISCOVER MY PASSIONS THROUGH TRAVEL, SO CAN YOU!

What are your passions? Do you already know, or are they waiting to be discovered?

Travel may also help you figure out and/or clarify your life's purpose.

What is your life's purpose? What will you stop at nothing to achieve? Do you already know, or is it waiting to be discovered?

I have found the following to be what I perceive to be some of the purposes of my life:

- To show others the power of gaining and analyzing new viewpoints through travel

- To inspire others to communicate authentically and in deeper ways -- not only by becoming a better listener but also by learning to communicate effectively in new languages and through nonverbal communication

- To teach others the critical role that cultural understanding and adaptation plays in the harmonious cooperation and peaceful coexistence of all human beings

I have reached a turning point in my life. I have recognized the great need to share these 15+ years of lessons I have learned through travel experiences throughout the world in order to:

- Inspire a greater understanding of the benefits of travel

- Promote a better understanding of one another as human beings, especially across cultures

- Show you that an entire world of adventure and growth awaits

- Show you how you can unlock your true potential through travel

- Motivate you to become the best version of yourself

- Arrive one day, at a place where we no longer experience violence. Where we no longer experience war. Where we are all at peace

Ignorance can turn into fear. Fear can turn into hatred. And hatred can turn into violence. By better understanding people of other races, ethnic groups, cultures, and so on, we can eliminate ignorance, and therefore simultaneously eradicate the fear of the unknown. So by educating ourselves about those who are different from us, the likelihood of hatred and violence therefore becomes greatly reduced. **And the prospect of peace emerges as a tangible and attainable reality.**

If I can accomplish my goals, you can accomplish your goals too!

When we travel, we open ourselves up to new ideas, new perspectives, and new experiences. When we travel, we learn how to adapt ourselves to new environments. When we travel, we connect. We connect with others. We connect with nature. We connect with ourselves.

It is my sincere hope that through my experiences, you have gained a better understanding of the benefits and value involved in stepping outside of your comfort zone. It is my sincere hope that you have seen how your life can be transformed through travel.

If I can grow from gaining new cultural perspectives, so can you.

If I can pick up additional languages through my global journeys, so can you.

If I have been able to transform through travel, so can you!

From the bottom of my heart, thank you for taking the time to read my book. I hope you have been inspired. I hope it has left a positive imprint on you. I hope that your life has been and will continue to be affected in the most positive of ways because of it.

If you haven't started planning your dream trip yet, WHAT ARE YOU WAITING FOR?! GO DO IT!! GO CHANGE YOUR LIFE!! GO CHANGE THE WORLD!!

And although this is only just the beginning of our time together, I'll end with a rather traditional statement to formally conclude this journey you have so graciously decided to join me on:

THE END.

About the Author

Rob received both his Bachelor of Arts in Spanish and his Master of Business Administration from The University at Albany in New York. He is an energetic and inspirational individual who encourages others to succeed! A student of the world, Rob is multilingual and has traveled to over 50 countries to better understand diverse people and cultures. He has worked in public and private education in Argentina and Japan, including with Berlitz. He has also worked with organizations such as Bloomberg and Dwellworks. He is a certified cross-cultural trainer who is dedicated to sharing his knowledge to improve the lives of others. He fuses his multilingual and intercultural skills with his travel, educational, and corporate experiences to deliver travel and cross-cultural training and consulting solutions to organizations, global teams, and individuals worldwide. To learn more about Rob and his work, you can connect with him in the following ways:

www.robmaisel.com

www.linkedin.com/in/robertmaisel/

robert.w.maisel@gmail.com

www.ingramcontent.com/pod-product-compliance
Lightning Source LLC
Chambersburg PA
CBHW011203090426
42742CB00019B/3394